D1476307

Nature's Chemicals

Nature's Chemicals

The Natural Products that shaped our world

RICHARD FIRN

OXFORD
UNIVERSITY PRESS

Chemistry Library

CHEM

OXFORD
UNIVERSITY PRESS

Great Clarendon Street, Oxford OX2 6DP

Oxford University Press is a department of the University of Oxford.
It furthers the University's objective of excellence in research, scholarship,
and education by publishing worldwide in

Oxford New York

Auckland Cape Town Dar es Salaam Hong Kong Karachi
Kuala Lumpur Madrid Melbourne Mexico City Nairobi
New Delhi Shanghai Taipei Toronto

With offices in

Argentina Austria Brazil Chile Czech Republic France Greece
Guatemala Hungary Italy Japan Poland Portugal Singapore
South Korea Switzerland Thailand Turkey Ukraine Vietnam

Oxford is a registered trade mark of Oxford University Press
in the UK and in certain other countries

Published in the United States
by Oxford University Press Inc., New York

© Richard Firn, 2010

The moral rights of the author have been asserted
Database right Oxford University Press (maker)

First published 2010

All rights reserved. No part of this publication may be reproduced,
stored in a retrieval system, or transmitted, in any form or by any means,
without the prior permission in writing of Oxford University Press,
or as expressly permitted by law, or under terms agreed with the appropriate
reprographics rights organization. Enquiries concerning reproduction
outside the scope of the above should be sent to the Rights Department,
Oxford University Press, at the address above

You must not circulate this book in any other binding or cover
and you must impose the same condition on any acquirer

British Library Cataloguing in Publication Data
Data available

Library of Congress Cataloging in Publication Data
Data available

Typeset by Newgen Imaging Systems (P) Ltd, Chennai, India
Printed in Great Britain
on acid-free paper by the
MPG Books Group, Bodmin and King's Lynn

ISBN 978-0-19-956683-9

1 3 5 7 9 10 8 6 4 2

RR
7/8/10

QK
99
A1
F57
2010
CHEM

Preface

Science is facts; just as houses are made of stone, so is science made of facts; but a pile of stones is not a house, and a collection of facts is not necessarily science.

Jules Henri Poincaré (1854–1912) French mathematician.

This is a book about ideas. It is ideas that excite, not factual knowledge. Consequently this book aims to stimulate thinking in its readership and uses "facts" sparingly. I do not expect readers to agree with every part of my narrative, or even to be interested in every one of the diverse topics I discuss. But by trying to make readers aware of the fact that Natural Products have played a very large part in world history, that these substances punctuate the days of every reader and that the trade in Natural Products is a major part of the world's economy, I will try to justify my view that every well educated person should understand what Natural Products are and why they are so important and fascinating. I also hope to start a process of putting a fragmented subject back together again, making the subject more attractive to teachers and students. Years of teaching revealed to me that many students have been taught to "learn" facts but many have little understanding of the real nature of scientific endeavour. Consequently, maybe controversially, I end chapters with some thoughts about the nature of the scientific processes that have helped, or hindered, the development of our understanding of Natural Products.

There are many people who I should thank for their encouragement and support over the years. I thank all former colleagues and teachers who engaged me in productive debate. Some people might be unaware of their helpful role. For example Colin Jenner at the Waite Institute, Adelaide, amazed me for 2 years by asking simple critical questions after every seminar which made me radically reassess the information presented by the speaker. Bob Bandurski at Michigan State University was another person who unwittingly improved my capacity to think. While a post doc at the MSU/PRL I was expected to attend, and help teach, an advanced course on plant physiology/biochemistry. After the first lecture I gave, on plant hormones, Bob came up to me and volunteered a critique of my efforts. Too many facts, a lack of a critical analysis and no narrative built around theory. "The audience needs to be made to think, even if they don't accept your analysis". He was right. The talks I gave subsequently at the PRL became progressively more radical, as I sought to probe the weaknesses of theories with inconvenient facts. This became a theme of my later work on phototropism and gravitropism (where I challenged long accepted dogma). I could not help using the same approach when I strayed into the subject of this book—Natural Products.

Unlike several people who helped me learn how to think critically, Clive Jones knows well the debt I owe him. Clive was my first graduate student when I moved to the

Biology Department at the University of York. Clive arrived with his own ideas about what he would work on—the Natural Products (phytoecdysones) in the bracken fern which supposedly protected the plant against insect herbivory. The inconvenient fact that Clive discovered was that the concentrations of phytoecdysones in bracken were too low to be of significance to the survivorship of the insect eating the plant. So why were these Natural Products made? This question lingered in our brains until some years later when Clive came back to York on sabbatical to work with John Lawton. Some time before Clive returned, I realised that one piece of information I had known for some years, seemed to have escaped us—specific, potent biological activity is a rare property for a molecule to possess. When I put that thought to Clive, he excitedly began to construct, in the course of an evening, a new model to explain Natural Product diversity. A little later John Lawton was organising a Royal Society meeting on co-evolution and he invited Clive expose our ideas to scrutiny. The audience gave the new model a very frosty reception so for the next decade Clive and I collaborated to refine our ideas in the light of the continuing criticisms. Without Clive's energy and confidence I doubt if our ideas would have matured.

My thanks also to York colleague John Digby who kept our work on plant tropisms alive while I pondered and wrote about Natural Products. York colleagues David Hoare and Simon Hardy patiently answered ill formed questions from me. My wife Ulla Wiberg was even more patient. After years of watching me write about, and talk about, Natural Products she tolerated me devoting part of my retirement to this book. Ulla has generously promised to translate this work into English one day.

Finally Ian Sherman, Helen Eaton and Carol Bestley at Oxford University Press have been patient, tolerant, supportive, calm and professional.

I will end with a quote from that fine thinker, polemicist and radical Thomas Paine:

> *"Perhaps the sentiments contained in the following pages are not sufficiently fashionable to procure general favour; a long habit of not thinking a thing wrong gives it a superficial appearance of being right, and raises at first a formidable outcry in defence of custom. But the tumult soon subsides. Time makes more converts than reason.*
>
> Thomas Paine (1737–1809)

Contents

1

What Are Natural Products?

"When I use a word", Humpty Dumpty said, in rather a scornful tone, "it means just what I choose it to mean—neither more nor less."

—Lewis Carroll

Summary

All organisms are made up of chemicals. There is a common collection of several hundreds of substances that are produced by all living organisms. However, hundreds of thousands of different chemicals are also produced by plants and microbes, with each species producing its own characteristic mix.[1] This much larger class of chemicals, often called Natural Products (NPs), are clearly not essential for life, but their production must bring some benefit to the producer. The different NP composition of pears and apples, for example, explains why these two fruits taste different. The difference in NP composition of lemons and roses gives each species its characteristic smell. Even within a species, the NP composition can vary; hence, individual varieties of apples or pears can have distinctive flavours.

In Chapter 2, it is argued that NPs have dominated, and continue to dominate, world economic activity and that these chemicals influence the lives of every individual, every day. Consequently, because NPs are so important, one would expect every child to learn about them but they do not. Why? The answer to this question tells us something about how science is conducted and taught. Although science is firmly compartmentalised (at school, university, bookshop or library) into subject disciplines such as physics, chemistry, zoology and so forth, there are no logical boundaries between subject areas. Scientific boundaries, like national ones, are human constructs and are not always wisely drawn. The subject of NPs became fragmented in the nineteenth century when it was split between chemists and biologists. Neither group nurtured their part of the subject. Biologists fragmented their studies of NPs into more subdisciplines, and the chemists became fixated on the structures of the individual chemicals. It is as if the study of orchestral music gradually became the study of some chords or individual notes.

Natural Products (NPs) and natural products—spot the difference

How sad and frustrating that the first hurdle the author has to clear is to try to explain why a natural product differs from a Natural Product.

Most English-speaking people can readily give examples of products that they consider to be 'natural'. Common examples would be sugar, butter, wood, honey and cotton. These *natural* products are considered by many as being somehow different (healthier? safer? more environmentally friendly? better?) from *synthetic* products made by humans, such as plastics, petrol, oil, pesticides and pharmaceutical drugs. As advertising agencies have learned, the term *natural* carries with it many positive associations, and it is ideally suited for misuse. The fact that nearly all natural products are now highly processed and refined by humans before being presented to the consumer is overlooked. So the commonly used term 'natural product' is vague, open to too many interpretations and features very little in this book.

Unfortunately, the term 'natural product' also passed into scientific usage in the nineteenth century. Scientists in the English-speaking world started to use the term *Natural Product* (sadly not always capitalised) to classify a particular type of chemical made by plants or microbes. At that time scientists were beginning to analyse the chemicals found in different plant species. These analyses revealed that some chemicals (many simple sugars, amino acids, some organic acids) were common to most plants studied. However, every individual plant species also contained a few chemicals that were distinctive, substances that gave the species its characteristic smell or taste. These distinctive chemicals were placed in the general category of Natural Products. So we have this Alice in Wonderland situation where all Natural Products are natural products but not all natural products are Natural Products (Figure 1.1). This was indeed an inauspicious, careless start to the building of a new area of scientific study. Many scientists have been unhappy with the term Natural Product and have suggested alternatives, the commonest being *Secondary Metabolite* (the reasons why this term was proposed,

natural products	natural product (containing a Natural Product)	Natural Product (NP)
Sugar	Tea (caffeine)	Penicillin
Starch	Coffee (caffeine)	Taxol
Wool	Cocoa (theobromine)	Vincristine
Wood	Tobacco (nicotine)	Cyclosporin

Figure 1.1. What is the difference between a natural product and a Natural Product? To chemists the answer was simple. When a chemist uses the term Natural Product (or sadly, even 'natural product' without capitalisation), they refer to a naturally synthesised substance that is not made by all organisms but is made by only a few species. However, the term 'natural product' is now much more widely used by the rest of society to mean anything not manufactured. This ambiguity is very unhelpful; consequently, this book will adopt the term NP for Natural Product, the term NP being chosen for its historical association and language independence.

and the reasons why this term is best consigned to history is explained later). Sadly, no alternative term for Natural Products has gained universal acceptance; hence the author feels that it is time to rid ourselves of this unfortunate legacy. In society and science, abbreviated terms are now widely used. Abbreviations have the advantage of being more easily adopted in other languages, yet they can act as an appropriate link to the past usage. Consequently, it is suggested that readers accept the abbreviated term NPs for Natural Products so that the confusion about the name can be left behind us before we start to explore the subject.

So, at this stage of our journey, the reader needs only to know that NPs are the distinctive chemicals which characterise particular plant and microbial species. Each species of plant produces its own mixture of maybe a few hundred NPs. The difference in NP composition explains why a pear tastes different from an apple or a rose smells different from a carnation. In the plant kingdom, hundreds of thousands of different NPs are being produced. Every human experiences many of these substances daily in the scents and flavours in their food and drink, but they are blissfully unaware of the many more chemically complex NPs that they ingest daily but are not sensed by the human sensing systems. Several chemicals which are well known to most citizens, because they are hugely prized, are NPs—morphine, caffeine, quinine, penicillin, the anticancer drug taxol, the drug vinblastine and so forth. With this brief summary of what NPs are, the reader can move on to subsequent chapters to follow their own interests or they can find out why the science of NPs has progressed so hesitantly in the past.

NPs—the subject forms and then fragments

Chemicals from minerals and chemicals from organisms—inorganic and organic chemistry

At the end of the eighteenth century, chemistry was emerging as an academic subject. It drew on two distinct streams of information from past human endeavours—the study of minerals and plant products. Knowledge of mineral rocks, and the metals that could be obtained from them, had been increasing slowly for thousands of years and was highly valued in all cultures. Likewise, the human experience in the production and processing of plant-derived substances to give food, dyes, fabrics and drugs (both pharmaceutical and recreational) was central to all human civilisations. Within all societies, a few people inevitably 'experimented' with new ways of processing minerals or plant products, guided by thoughts of how to produce something of greater value from the starting substance. Since the days of the ancient Greek philosophers,[2] it had been argued that most substances, whether mineral or derived from plant or animal sources, were made up of other entities. Over the centuries, experimental manipulations, guided by theoretical or empirical motives, had been undertaken to try to separate and identify these entities. In the eighteenth century, deductive reasoning was increasingly used to devise general rules that might be applied to the understanding and manipulation of minerals. By the beginning of the nineteenth century, such studies, undertaken particularly intensively by Swedish and Finnish chemists, had led to the identification of 36 elements (carbon, hydrogen, oxygen, sulphur, nitrogen, etc.). By 1830, the number of known elements had risen to 53, with almost half the elements now known being discovered. The contemporaries studying the chemical properties of plant or animal products knew of old evidence that natural materials might also be composed of simple substances combined in some way. For example, distillation had been widely used since the tenth century to separate alcohol from fermented broths and to produce oil of turpentine from pine resin. A significant advance was made in the years 1769–1785 when the Swedish chemist Scheele isolated a number of different sour substances in plant- or animal-derived products—tartaric acid from grapes, citric acid in citrus, oxalic acid in wood sorrel, malic acid from apples, gallic acid in galls, lactic acid in sour milk and uric acid in urine. Scheele deduced that

- organisms contained many different substances,
- the amount of any one substance could differ in different organisms,
- different substances could share some properties (e.g., there were many substances that humans perceived on the tongue as sour).

Scheele did not possess the ability to identify the elemental composition or the structures of the substances he had isolated, but others soon developed the techniques needed for the crucial next stage. The French chemist Lavoisier's studies of combustion (1772–7) produced a method of determining the elemental composition of most chemicals made by organisms. When Lavoisier (aided very considerably by his wife, Marie-Anne Pierrette Paulze) burned a natural substance (e.g., sugar, ethyl alcohol and acetic acid) in oxygen (discovered first by Scheele in 1772 and then independently again by Priestley in 1774), he found that the only combustion products were carbon dioxide and water. He deduced that such natural substances contained carbon, hydrogen and, possibly, oxygen. By quantification of both the amounts of the starting materials and the amounts of the products formed by combustion, Lavoisier could determine the relative amounts of each element in the combusted product. Subsequent studies showed that some other chemicals made by organisms (e.g., urea, hippuric acid, morphine) produced carbon dioxide, water and nitrogen when burned in oxygen; hence they must have contained nitrogen as well as carbon and hydrogen. Combustion in oxygen was widely adopted as a simple way of determining the elemental composition of all chemicals that were found in organisms. Soon it was discovered that different substances could share the same elemental composition, and it was proposed that the way in which the carbon, hydrogen and oxygen atoms were joined together must be characteristic of each substance. After the concept of valency[3] was accepted, the location of the carbon–carbon links (bonds) was seen as the key to identifying each chemical, and the concept of the *carbon skeleton* emerged. A convention as to how such *chemical structures* could be drawn on paper was adopted and the scene was set for the characterisation of the structures of all naturally made chemicals (Figure 1.2).

It was at this stage of the development of chemistry that the problems of terminology began. In 1807, the influential Swedish chemist Berzelius had proposed

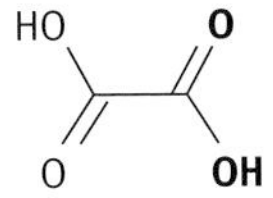

Oxalic acid $C_2H_2O_4$

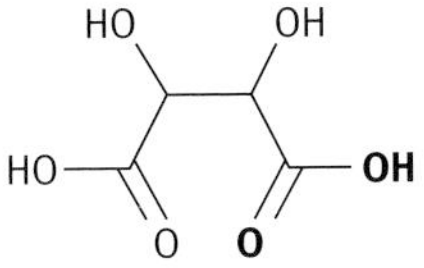

Tartaric acid $C_4H_6O_6$

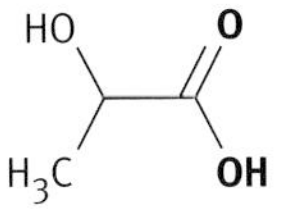

Lactic acid $C_3H_6O_3$

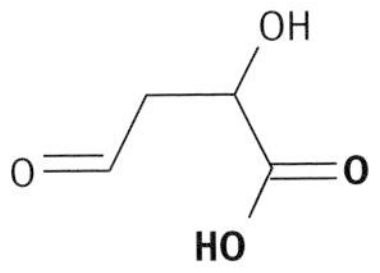

Malic acid $C_4H_6O_4$

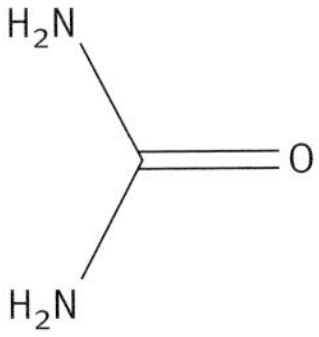

Urea CH_4N_2O

Figure 1.2. Scheele isolated a family of naturally occurring sour substances. Subsequently, the elemental composition of each substance was determined using Lavoisier's combustion method and later the different structures were proposed. Each member contains common structural element, a carboxylic acid group which gives each its sour taste. The two-dimensional representation of chemical structures as shown is convenient but can be misleading. Also shown is the structure of urea, the first naturally occurring substance to be made in the laboratory by Wohler (shown), who provided the first experimental challenge to the concept of vitalism.

that the chemistry of substances made by living organisms should be referred to as *organic chemistry* to distinguish it from the chemical studies of mineral or inorganic substances—*inorganic chemistry*. He reasoned that in contrast to inorganic substances, such as mineral ores or metals, organic substances were sensitive, even delicate, as evidenced by their ease of combustion. In particular, he noted that chemists could make inorganic substances, such as various salts, but he argued that chemists could never make organic substances—these substances could only be made by living organisms. Berzelius believed and argued that only living organisms possessed the *vital force* needed to make these delicate organic compounds. At that time the merging of scientific ideas with what we would now regard as a mystical concept such as vitalism was in no way remarkable and Berzelius's view was widely accepted.

The vital force—science to anti-science

In 1828, the young German chemist Wöhler, who had studied in Stockholm with Berzelius, showed that Berzelius's postulate, that humans could not make organic molecules, was wrong. Experimenting with the inorganic substance ammonium cyanate, Wöhler made the organic substance urea (Figure 1.2). After repeating the experiment many times, Wöhler finally wrote to Berzelius stating, 'I must tell you that I can prepare urea without requiring a kidney or animal, either man or dog.' As often found in science, this simple disproof of the existence of Berzelius's vital force was not immediately or universally accepted. Berzelius was a scientist of great authority, while Wöhler was only starting to make a reputation. Vitalism, supported by authority, was given the benefit of the doubt and the concept lingered on and eventually took on a new life. However, the majority of the new generation of organic chemists did not need the concept of vitalism to guide them in their attempts to make new chemical structures or to determine the structure of chemicals isolated from plants. However, a new subgroup that emerged from organic chemistry, the *physiological chemists*, found a new role for the concept of vitalism and, in doing so, broke away from the organic chemists.

Physiological chemists (known as biochemists since the twentieth century) were those chemists who were interested in the chemical transformations that are taking place within organisms—ordinary organic chemists were more interested in the chemical transformations that they could make happen in their flasks, beakers or retorts. Physiological chemists were inspired by the work of Persoz and Payen. In 1833, these two French chemists had reported that an extract of malt (the brown, sweet mixture of sugars derived from the germinating grains of barley) in water could break down insoluble starch to produce soluble sugars. They had discovered the catalytic proteins, later to be called enzymes, which were thought to be unique to living cells. The new slant given to vitalism was that only whole living cells were thought to carry out enzymic transformations. Even Wöhler subscribed to this view as did the very influential German chemist Liebig. The concept of vitalism lived on in this new form until 1897, when Buchner found, by accident, that enzymic activity could be measured in extracts of yeast cells which lacked any living cells.

However, even after all traces of Berzelius's ideas on vitalism had been disproved by experiment, his associated term 'organic' survived.[4] After Wöhler had shown that humans as well as organisms could make 'organic' chemicals, the term *organic chemistry* became the chemical study of any carbon-based compounds, and that usage remains.

Organic chemistry as a subject fragments—synthetic and Natural Products

As the subject of organic chemistry grew during the second half of the nineteenth century, specialisms emerged within that subject. Some organic chemists were mainly interested in the generation of new carbon-based molecules (organic); consequently, they developed new or improved methods of synthesising known and novel organic molecules. This branch of organic chemistry became known as *Synthetic Organic Chemistry*. Other organic chemists were fascinated by the diversity of chemical structures made by organisms, and these chemists developed methods to isolate, purify, characterise and synthesise such naturally occurring substances which they called, in the English-speaking world, *Natural Products* to distinguish them from the synthetic chemicals of interest to other organic chemists; hence the birth of *Natural Products Chemistry* and the common usage of the term of Natural Products in the English-speaking world. However, Natural Products Chemistry was soon to fragment further.

As the chemical composition and structures of substances derived from living material was explored, it became clear that many substances were found in the majority of organisms but some substances seemed to occur only in a few species. For example, the 22 common amino acids, glycerol, sucrose or glucose could be found in most animals, plants and fungi. But everyone knew that many plant species produced chemicals, such as scents, that were characteristic of that species—roses, mint, pepper, cloves and so forth. Not surprisingly, by the mid-nineteenth century, the most widely distributed Natural Products, those common to most organisms, were attracting most interest. The chemistry of digestion and assimilation, catalysed by the newly discovered enzymes, was an exciting field of study and the Physiological Chemists increasingly concentrated on these universally important substances. Step by step the sequence of the enzymic conversions which led to the formation of each substance was established. Each new enzymic step was added to a 'map' of the inter-conversion routes of the chemicals common to most organisms. As the decades passed, the Physiological Chemists studying the enzymes in these widely shared pathways had less and less in common with those Natural Products Chemists studying the chemical structures of the rarer strange chemicals, each of which occurred in only a few species. The techniques and concepts used by those studying the rare, exotic chemicals had much more in common with Organic Chemists than Physiological Chemists. Inevitably, the subject of Natural Product Chemistry fragmented. By the early twentieth century, some institutions had created new Physiological Chemistry Departments (or Biochemistry Departments as they were subsequently called) to work alongside the Chemistry Departments. In general, the new Biochemistry Departments concentrated on the biochemistry that was nearly universal to all organisms; consequently, the Natural Product Chemists were usually left in Chemistry Departments where they were more at home. Thus the scope of the subject of Natural Products Chemistry, which had started in the nineteenth century as a study of all naturally produced chemicals, was reduced in the twentieth century to the study

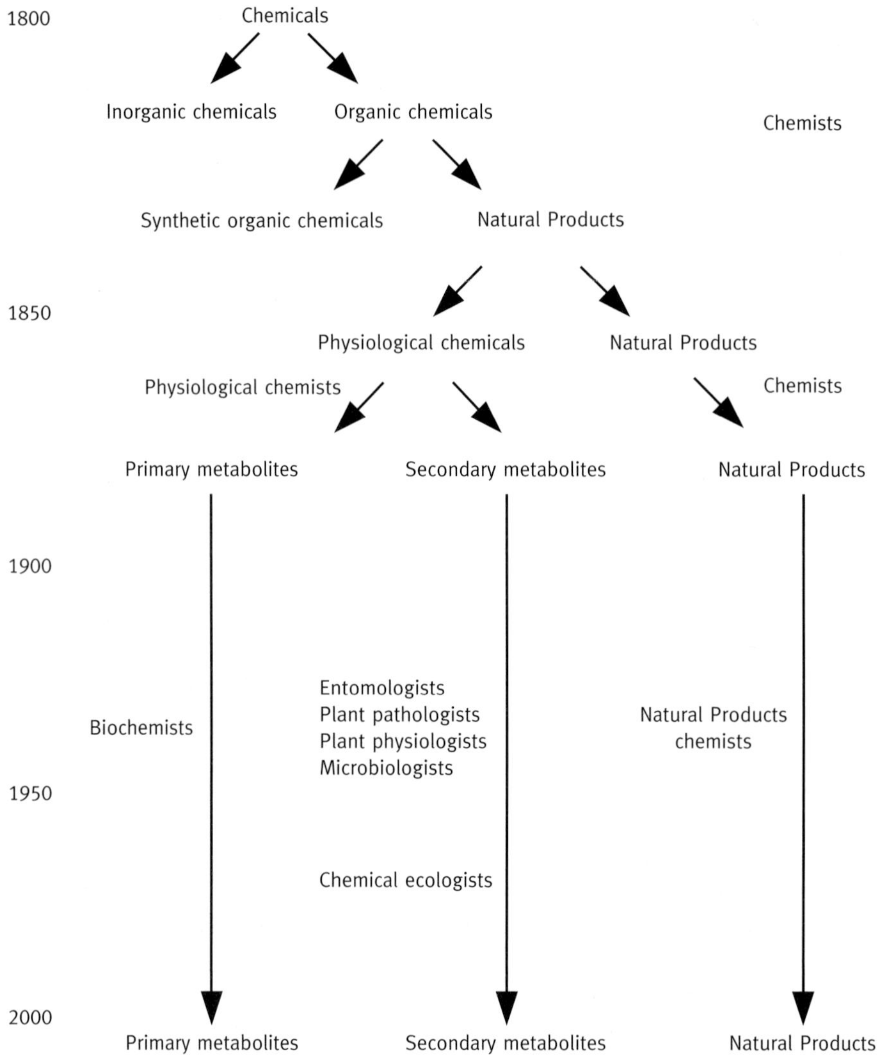

Figure 1.3. The time scale of fragmentation of the study of NPs—a fragmentation that has resulted in the current unsatisfactory situation where different groups of scientists use their own names for the same class of substances.

of only those chemicals which were often found only in a few organisms. To a chemist, a Natural Product is simply a fascinating naturally occurring carbon-based chemical that is of little interest to most biochemists. To most biochemists, Natural Products are occasional useful tools (because a few have a very powerful ability to perturb cell or enzyme functioning) but not much to do with 'real' biochemistry—that is why the subject was left to the chemists.

To further complicate matters, in 1891, the German Physiological Chemist Albrecht Kössel (who won the 1910 Nobel Prize for Medicine) unknowingly aided this fission of the subject of metabolism when he proposed that plants had two distinct types of metabolism, 'primary' and 'secondary'. He proposed that *primary metabolites* were involved in basic processes of the cell and were common to many organisms. *Secondary metabolites* were made by distinct pathways limited to some organisms; hence they served a less vital role. Kössel's *primary metabolism* was the subject taken over by biochemists, and his secondary metabolites were the Natural Products being studied by chemists. A consequence of Kössel's categorisation is that the terms *Natural Products* and *Secondary Metabolites* are still used synonymously (Figure 1.3). The former term is used mainly by chemists and the latter mainly by scientists with a biological background. Could there be a better example of the fragmentation of a scientific subject than one where two different groups have the different names for the same subject?

Time to 'rebrand'?[5]

The result of this century-old fragmentation of the subject is well illustrated when looking for information about *Natural Products* or *Secondary Metabolism* in current biochemistry textbooks. When the indices of 10 important current biochemistry textbooks were examined in 2000, none contained the words *Secondary Product*, *Secondary Metabolism* or *Natural Product*! While it is understandable that biochemists should concentrate on the few hundred chemicals that are commonly produced by most cells, a sense of fair play or balance would surely demand that some reference be made to the fact that most of the world's biochemical diversity resides in another type of metabolism.

Many scientists studying *Natural Products* or *Secondary Metabolites* have been participants in debates aimed at 'rebranding' their subject but without success. Debates have been held as to which is the preferable term and new terms have been suggested, such as *semiochemicals*. A consistent theme in such debates has been the feeling that the use of the term 'secondary' has been unfortunate because there is an implication that such compounds are not really important. The term semiochemical (from the Greek word *semeon*, a signal) was inspired by the belief that Natural Products were involved in signalling, but the term tends to be used only by those interested in chemical signalling between some groups of organisms (e.g., between insects—see Chapter 8) but the term is inappropriate because most Natural Products are not involved in signalling. The decision by the author to use the term NP in this book, instead of Natural Product, was taken on the basis of historic precedence, but also because it will be argued in Chapter 9 that Kössel's alternative term, Secondary Metabolites, was unhelpful and wrong!

Which organisms make NPs?

Although microbes and plants are the organisms that are most commonly exploited for their NPs, it is very hard to draw clear boundaries between organisms that do and do not make NPs. For example, humans make some substances of unknown function.

However, because digestion and excretion was central to the developing subject of Physiological Chemistry in the nineteenth century, such reactions were considered to be more closely related to 'normal' metabolism than NP chemistry; hence the role of animals in generating chemical diversity was largely ignored by those interested in NPs. Such a view makes these 'degradation compounds', typically formed in the liver, different from NPs. However, it is now clear that this approach is unhelpful and a broader, more holistic approach will be more productive. There are useful analogies between processes occurring in the liver and processes found in many plants and microbes, and, in later chapters, it will be argued that such links are not fortuitous but a consequence of the fact that organisms have evolved to make and adapt to NPs. However, the chapters that follow will concentrate on the remarkable NPs that plants and microbes make because it is those chemicals that have influenced the history of humans and which continue to drive the world economy.

What does this chapter tell us about the way science works?

Clearly the way in which the study of NPs developed in the nineteenth century has had a profound effect on the study of NPs in the subsequent centuries. For psychological reasons, most people like to gather with people who share their view of the world, but whenever such groups form, schisms always develop and subgroups drift off on their own. When such splits occur, some of the ideas shared with the larger group are retained but the new group always has to have a novel perspective. In this type of group behaviour, scientists are no different from those with religious beliefs. While scientists like to think their groupings are logical, based on some accepted realities, all scientific disciplines are human constructs. Those who form a new scientific discipline establish a way of thinking which persists because those who come to study that discipline are now taught what is virtually a creed. The common ideas that united a larger subject are given less emphasis than the key ideas of interest to the smaller group. With the benefit of hindsight, it is clear that when the study of NPs emerged from the study of chemistry in general, the most important idea needed to understand NPs was simply not one that chemists were seriously interested in—evolution. Maybe because the biological concept of vitalism had not proved productive, when vitalism was finally discarded, the biological concept waiting in the wings, selection based on natural selection, was not seized to provide a conceptual framework to shape the thinking of those interested in NPs. Yet, like all biological phenomena, NPs can only be understood in their evolutionary context. The laws of chemistry and the laws of physics apply to individual NPs, but these laws by themselves cannot provide any perspective on why NPs exist, what they do and why humans are in thrall to these chemicals. It is significant that within the broad subject of biology the study of NPs was closely linked to evolutionary ideas in some areas and not in others. Biologists closest to chemistry, the biochemists, were the least influenced by evolutionary thinking, while those working on NPs and their ecological role were working within a clear evolutionary framework. However, any interaction between organisms occurring at population level are really the result of processes occurring in individuals, and within the individual the processes are about the interactions of molecules, which themselves depend on molecular structures and the laws of physics. Any attempt to understand the processes that occur in any hierarchy must be built on an understanding of all the processes occurring at the lowest levels, because everything at the higher levels is governed by the laws that have shaped the outcomes of processes at the lowest levels. Sadly, the few simple rules that govern the natural world tend to get obscured by the increasing large piles of 'facts'[6] and nowhere is this better illustrated than in the study of NPs. Traditional books about NPs tend to be full of details (the chemical structures of thousands of substances or the subtle interactions between thousands of organisms), so it is not a lack of 'facts' that stops non-NP enthusiasts appreciating them. Consequently, this book tries an alternative approach, bringing to the fore a few simples rules and being very sparing with the 'facts'.

2

The Importance of NPs in Human Affairs

A historian who would convey the truth must lie. Often he must enlarge the truth by diameters, otherwise his reader would not be able to see it.

—*Mark Twain*

Summary

Humans often value a rare object more than a common object, even if the rare object has little practical value. History suggests that this generality applies to some NPs. The rarity of a particular NP at any one time is a consequence of the fact that each NP is usually made by a few closely related species (for reasons that are explored in Chapter 5). If individuals of such an NP-rich species are quite rare, then the unique mix of NPs characteristic of that species will be in limited supply. If humans find that mix of NPs attractive, then opportunities will arise for those who control the supply of the plant, or one of its NP-rich products, to make very large amounts of money. With money comes power and power dictates world history. Hence, it is hardly surprising that the human obsession for certain NPs (e.g., coffee, tea, spices, morphine, cocaine, hops, quinine, etc.) has influenced world history. However, if one looks at NPs as a group rather than each individually, the importance of NPs in human history is even more striking. Indeed, it is apparent that over the centuries, those rulers who have gained a near monopoly on the supply of certain NP-rich materials have shaped the current geopolitical map of the world. The wealth created by trading in NPs was used to build great cities with magnificent houses and public buildings, cities where arts and science flourished. Even today the legal, and illegal, trading in a few NPs still dominates the economic activity in the world (Figure 2.1). Remarkably, this very obvious fact is rarely appreciated.

Making money from NPs

Simple market economics

Humans in all the ancient cultures selected appropriate plants to produce the carbohydrates, fats or oils and proteins needed for their sustenance. Most of the chosen crops produced structures (e.g., seeds, tubers, etc.) that the plant had evolved for propagation

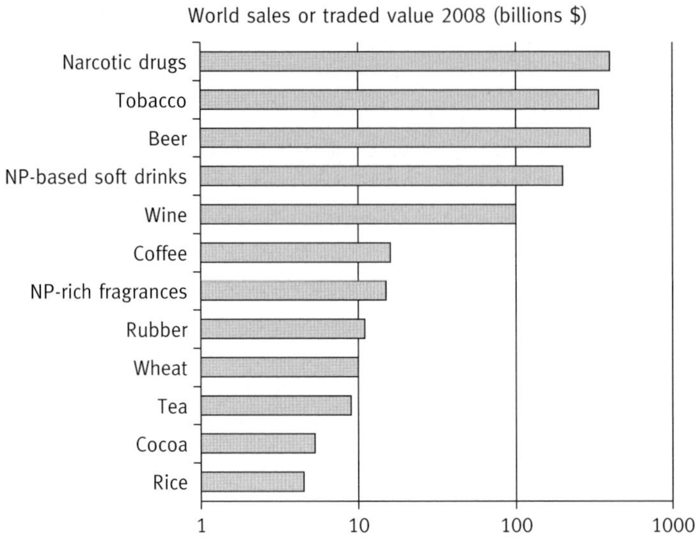

Figure 2.1. The world sales or trade values of some major plant products. It is doubtful whether the figure for each is strictly comparable because the monetary value of a product depends on several factors—the amount of processing, the ease of maintaining a monopoly, whether taxes are included, and so on. For example, the monetary value of narcotic drugs increases many fold from grower to consumer, much more than the increase in the value of wheat for example. Because many countries grow wheat, most of which is consumed domestically, the total value of the world wheat sales ($120 billion in 2008) greatly exceeds the amount traded. In contrast, many NP-rich plant products are grown mainly for export, hence have greater significance in world trade. The figures are mainly derived from FAO data covering the years 2005–8.

purposes. Evolution has ensured that these structures contained the appropriate mix of elements needed to sustain the crucial early life of the next generation of the plant and that these substances were stable and hence stored well. Many organisms, including humans, have evolved to exploit the storage reserves of plants. One of the criteria for the selection of domesticated crops was that the plant storage substances were contained in high amounts in an easily harvested and stored structure and that semipurified palatable substances could be produced with relatively little effort using simple techniques. For example, most grains contain large amounts of starch and the grinding of the seeds can produce 'flour' enriched in starch. However, such processed grains have a low odour and flavour because they contain few NPs (could it be that basing ones diet on high odour, high taste staples could burden the body's NP processing systems? See Chapter 7). Given the dependence of humans on their food crops, it is understandable why the main plant products traded in the world, in terms of tonnage, are grains, sugar, oil and protein containing seeds. These are *commodity products* containing chemicals of relatively low value (except in times of famine).

Because all plants possess the capacity to produce storage carbohydrates, storage lipids (fats and oils) and storage proteins, it is relatively easy for any group of humans to find some local plants that can supply these substances. The domesticated grasses (rice, wheat, oats, barley, rye, maize) were not only attractive to grow, harvest, store, transport and process, but also capable of yielding reliable harvests when grown over a range of soils and climates. The major grain crops such as wheat and maize are now grown on every continent. Because wheat starch is similar to barley or maize starch, it is possible to substitute one form of starch for another in several uses—brewing, the production of animal food, biofuel production or the production of sweeteners. Finally, the ability to store dried grain for many years means that surplus production in one year, in one part of the world, can benefit consumers in subsequent years in parts of the world where local grain production has failed to meet demand. The fact that farmers all over the world can produce starchy grains, and the fact that it is easy to substitute one starch for another for some uses, means that highly competitive markets have kept grain prices low for decades. The same arguments apply to other commodity food products such as plant proteins, plant oils and plant fats where competition and substitutability keep prices low. So a general lesson emerges. Throughout history, except in times of famine or very limited supply of staple foods, wealthy societies place rather little value on their staple foods. In particular, the wealthy members of society never spend much of their income on staple foods.

In contrast to the commodity products, the NP-rich plant materials added to the staple ingredients to give flavour, odour or colour to a food have often had very high value. It is the NPs that humans desire or even crave in their foods, not the nutritional substances. Not only have humans appreciated NPs in their foods, they have also used NPs much more widely. Drinks are more enjoyable if they have a flavour. Life becomes more pleasant if the nose is stimulated by the NPs in scents or perfumes. Pills and potions containing NPs, NPs that might have a pharmacological effect, are also more likely to benefit from a placebo effect if they have a distinctive taste. The most extreme case of this interest and attraction to certain NPs is found in narcotics.

But why are NP-rich plants more expensive than commodity or crops? Simply because, in contrast to the easily substitutable staple foods, the plants that produce highly attractive NPs usually have a much more limited geographical distribution. Consequently, there is less competition in the market place and substitution remains very difficult in most cases (e.g., there is no satisfactory synthetic coffee, tea or chocolate). Some of the NPs used in scents and flavours have been substituted with synthetic chemicals but even then many consumers were prepared to pay a premium for plant-derived flavouring (e.g., natural vs. synthetic vanilla).

The difference between commodity products and NP-rich plant products are summarised in Table 2.1. The properties of sought after NP plants show the factors that helped growers and traders establish a monopoly of supply. When each of those factors is examined it becomes clear why for every desirable NP, the monopoly is finally broken.

Table 2.1. The key differences in the characteristics of crops producing commodity products and crops producing NPs.

	Commodity crops	NP crops or products
Geographical range	Very wide	Very limited
Stability in storage	Very good	Mostly poor
Substitutability	Very high	Very low
Ease of transport	Increasingly economic	Always economic
Price per kg	Very low	Very high

Geographical range

By increasing the geographical range of any valued NP producing species, the monopoly enjoyed by the original producers, more particularly the traders who controlled the routes between growers and the consumers, was eroded.

- Improved husbandry skills allow NP-rich plants to be propagated and grown in more geographical regions, hence increasing supply (see sections on coffee, tea, vines, hops, rubber, etc.).
- Variants can usually be found within a population of plants that can survive in environments that were once considered unsuitable and further selections eventually produces plants that can produce a profitable harvest.
- Navigational techniques and ship technologies, which enabled transglobal voyages to begin in the fifteenth century, led to an accumulation of knowledge of the existence of equivalent climates, soils and weather and this knowledge encouraged people to identify certain valuable NP-rich plant species that might grow well (see section on spices).

Stability in storage and distribution

Because products rich in desirable NPs command a high price, it becomes economic to build storage facilities that reduce NP loss and to invest in improved packaging—look at foil wrapped tea or vacuum wrapped coffee. Freeze drying, controlled atmosphere packaging and optimum harvesting also help deliver the best NP-rich products to the consumer.

Substitutability

There are examples of three kinds of substitutability:

- Species substitution, where one species that produces the sought after NPs is replaced by another similar species that produces a similar product—for example, coffee and tea.
- Chemical substitution, where chemists devise synthetic routes to make commercial production of the chemical economical—vanillin is one example.

- Combined species substitution and chemical synthesis—the best example of this is the production of analogues of the anticancer drug taxol which is now made by using a species related to that which makes taxol (see Chapter 7). This related species make no NPs with useful anticancer properties, but one of its NPs can be converted to the useful analogue of taxol by chemists.

Ease of transport

Once again the high value of NPs relative to their weight enabled the investment in special ships (e.g., the beautiful nineteenth-century tea clippers) and facilitates such craziness as the airfreighting of roses (rich in NPs that smell nice and have pretty colours) from South America or Africa to Europe.

Adding value—miracle cures, patents and branding

Although many plants valued for their NP content were of limited geographical distribution, some others, especially annuals, were less demanding in their cultural requirements hence were more widely grown. This was especially true in the case of medicinal herbs. For medicinal use, many herbs (mint, thyme, feverfew, camomile, etc.) are needed only in small amounts and, given considerable care, can be grown over wide geographical areas (herb gardens were a common feature of monasteries throughout Europe). Many herbs can be propagated easily, either by seed or vegetatively, and hence are available to even the poor in many societies. However, extra value could be added to these materials by the *herbalist* in a way that was impossible for the miller and the baker processing a grain. Humans will pay handsomely for the promise of health but meagrely for the promise of food, unless they are starving. Most individuals know from experience that many different foods will satisfy their hunger but they are also likely to believe that only one herbal potion will treat their ailment. Consequently, even when herbs were actually widely available to many people, a few people found a way of 'adding value' using the NP content as the selling point. A mixture of substances, including NPs to give colour, odour and flavour, could be sold to people suffering a wide range of different ailments. The herbalist could make money by taking cheap substances and combining them in a supposedly unique way. The skill of the herbalist was valued in most human societies for thousands of years and even remains appreciated today.

In the nineteenth and early twentieth century, the developed world gradually replaced locally made herbal remedies with more purified, processed NPs which commanded even higher prices because patent laws could restrict the competition in the sales of such products. However, even NPs which were not protected by patent were to benefit from a remarkable form of monopoly that began to infect the developed world in the nineteenth century and spread worldwide in the twentieth century—the monopoly of the brand. Indeed, some of the world's first major brands were NP-rich products (see section on tobacco). A nice example of this is found in cola drinks where simple, cheap ingredients, including several NPs, are sold to gullible consumers who seem prepared to pay extra money simply to drink brand 'A' rather than drinking the equally refreshing unbranded competitor. Another example, the mixture of NPs that we call coffee, is sold in branded cafes at a vastly inflated price, the consumer effectively awarding the retailer a form of monopoly. The most extreme version of the exploitation of the human obsessions with NPs and brands is the perfume market—as I write this I find one perfume on sale for £230,000 per bottle! One hopes that there were no buyers but the 2006 world sales of less expensive fragrances were still a remarkable $16 billion.

The history of the human obsession with NPs

The account that follows comes from an English-speaking north European with no training in history or economic history. The narrative will attempt to illustrate the way in which a few NPs have influenced human economic activity.

Ancient history

The characteristic taste or smell of a food is used to signal an acceptable or non-acceptable food source. The evolution of the senses capable of detecting NP has long predated humans. For example, it is clear that many insect species select their food source, or the food source for their offspring (i.e., which plant is chosen to lay eggs on), partly on the basis of the NP composition (see Chapter 9).

The discovery of fennel, cumin and coriander seeds at some ancient burial sites suggests that taste and/or smell was incorporated into human cultural practices very long ago. A few cloves in a charred vessel found in a settlement on the banks of the Euphrates in Syria have been dated to about 1700 BC and because cloves grew thousands of miles to the east in the Spice Islands, this suggests that NP-rich products were being trade over very long distances at an early stage in human history.[1]

The Indo-Iranian cultures

The Indo-Iranian culture, which archaeological evidence suggests stretches back beyond 2500 BC, was centred initially near the Ural River and then it spread to cover large areas of the Eurasian steppes, stretching down into what is now Pakistan and northern India. During a later period lasting from 1800 BC to 1000 BC there developed a 'soma culture'. Soma was a drink supposedly favoured by the gods and those who consumed it supposedly experienced increased power and euphoria. A debate continues as to which plant species (or indeed fungal species) provided the NP, which was the active ingredient of the drink. One candidate is a species of the genus *Ephedra*, which contains Ephedrine. The concept of the power of soma is also evident in Hinduism and in some western literature (most notably Aldous Huxley's *Brave New World*).

Mediterranean cultures—the demand for NPs grows

The pre-Roman spice era

The history of the cultures that participated in trade around the Mediterranean suggests that the consumption of NP-rich spices spread west and north, so the ports at the eastern end of the Mediterranean grew rich on the profits from the trade in NP-rich products.[2] For hundred of years, the Egyptians used their position on the Red Sea to trade by sea with India via Arabian merchants in Yemen. The ancient Egyptians had been using NP-rich products like balsam and myrrh, imported from the Yemen kingdom in south Arabia, in embalming since the third millennium BC.[2] The numerous conflicts in the region during these times can be interpreted as being caused by a desire to gain control over the trade in NP-rich products. However, as the demand for these products increased, more northerly trade routes were developed so the wealth from NP trading spread to several rulers controlling the Mediterranean ports, and the trade routes leading to them, in the Middle East (Figure 2.2). From 500 BC, the Greeks were importing cinnamon and pepper from India and nutmeg, ginger and mace from the Spice Islands in the Far East (islands which are now part of Indonesia). The use of these spices and others in medicine, religious practices and cooking spread throughout the Greco-Roman Empire

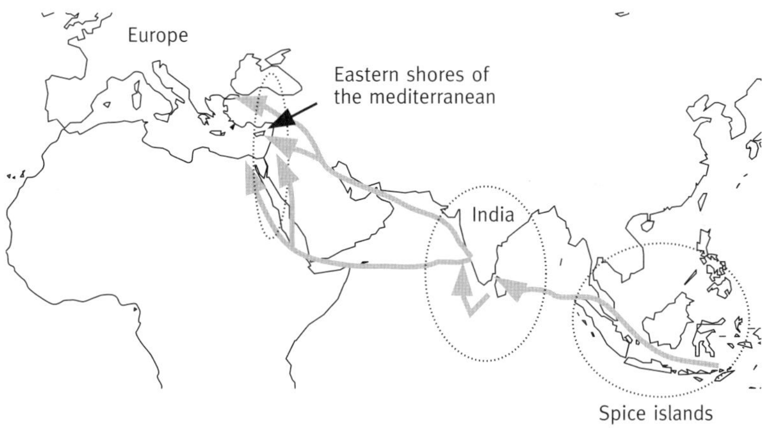

Figure 2.2. The major trade routes leading to Europe from the regions of spice production before the end of the fifteenth century, after which some western European nations (Portugal, Spain, The Netherlands and England) opened up ocean routes via Africa or South America. The control of the major sea ports at the eastern end of the Mediterranean and land routes across the deserts of the Middle East leading to them were key to capitalising on the spice trade. By the sixteenth century, when it was possible to bypass the Middle East traders, controlling the trade in India became important to maximising profits from spice trading. A little later the few major European nations seeking to control the spice trade fought for control of the areas of spice production in the Far East.

after Rome conquered the Greeks around 150 BC; subsequently, the use of spices spread into other parts of Europe as the Roman Empire extended west and north.

The Roman spice era

Given the growing importance of spices in Roman culture it is hardly surprising that the Romans sought to take control of the supply routes of these products from the East. The Romans annexed Egypt in 30 BC. Roman merchants invested in new ports on the Red Sea and they improved camel train routes across the desert to the Nile where they could load their precious cargo into barges for transport to Alexandria. From Alexandria the spices could be shipped to Rome and other ports in the Roman Empire. According to the geographer Strabo (63 BC–AD 24), 120 ships per year were engaged in the Roman trade to India and some of the freighters displaced 1000 tonnes (dwarfing by a very large margin the Portuguese, British or Dutch vessels that were to rediscover the Indian spice ports nearly 1500 years later). An understanding of the annual cycle of monsoons in the Indian Ocean gave these traders the ability to cross rapidly to their destinations. These huge vessels carried many different goods each way, with spices being one of the exotics that the Romans desired. This new spice trade flourished as the appetite for spices in the imperial city and the empire grew. The Romans found new uses for spices, especially pepper, in their increasingly rich and diverse cuisine. As the trade grew, the price of pepper fell to the extent that even soldiers serving at an outpost in northern England could expect some pepper to spice their rations. However, while the price of individual spices fell for a time, the overall cost of importing spices from India grew and this began to put a strain on the Roman economy. The problem for the Romans was that Indian merchants found rather few Roman goods attractive; hence, the Indians insisted on bullion in exchange for their spices. A Roman balance of trade problem, so familiar to modern economists, became evident quite quickly. Tiberius in AD 22 condemned the new obsession with imported luxuries before the Senate but many of his audience, no doubt, went home to feast themselves on their delicious spicy foods that evening. Not for the first or last time in history, would a rational argument against the unnecessary consumption of NPs fail to change the minds of those whose senses were enjoying them so much.

By AD 200 the price of spices had increased again to the extent that trade declined. As the power of Rome declined, the empire began to fray at the edges. The Red Sea ports fell to the Blemmyes, an African Nubian tribe who cut the direct Roman links with India. The spice trade between east and west was now to be dominated for nearly 1000 years by Arab and North African traders. However, even as the Roman Empire declined in Europe, the taste for spices continued. Indeed, some of Rome's enemies even sought spices as a tribute from Rome. Rome retreated from the west Mediterranean and mainland Europe and retrenched in the third century AD to the Eastern Roman Empire (or Byzantium) with its capital in Greek-speaking Constantinople (modern Istanbul). It is surely significant that Byzantium was at the end of one of the northern spice route and gave this last bastion of Roman power access to some spices that had not passed through the Red Sea.

The Arabian control of spices

The Byzantium and Jewish spice traders were to be increasingly marginalised by the remarkable growth and power of the Arab, and subsequently Muslim, traders. The Arabs had been very successful in gaining influence further to the east of the spice route—in the Indian subcontinent and beyond. There were Arab-speaking merchants in Sri Lanka by the fourth century AD and by the eighth century the Arabs even had their own commercial enclave on the Chinese mainland at Guangzhou (known formerly as Canton). However, the founding of Islam[3] forged a powerful set of links between wealth, power, culture and NPs. Mohammed (570–632 AD) brought his religious insights to many Arab tribes and united them under Islam. Mohammed married Khadija, the rich widow of a spice trader. Mohammed had worked briefly for Khadija, taking charge of one of her caravans taking goods to the Syrian ports. The favourable reports of his character and ability encouraged the older Khadija to subsequently marry Mohammed. Given Khadija's business as a trader in spices, it is not surprising that the new religion spread along the existing spice routes as well as locally. There was an Arab community established on the Malabar Coast of India 500 years before the Moghuls came from the north to bring Islam to other parts of that subcontinent. (The Indian coast was also the home to the oldest Jewish community outside the Middle East because of the involvement of Jews in the spice trade.) Because the spice route stretched to the country now known as Indonesia, Arab traders and their Islam religion became established in Indonesia. So the human quest for NPs explains why Indonesia was eventually to become the world's most populous Islamic country.

By AD 700, the Arabs had pushed the Byzantium Empire back to an area around current Turkey and the Byzantium influence in Europe was confined to a few enclaves (including Rome, Naples and some ports on the northern Adriatic coast). However, this limited access to the European trade routes enabled Byzantium to continue its profitable spice trading to mainland Europe until it lost control of its European ports in the eighth century. Byzantium lost trade into Europe to a community of fishermen living on some easily defended marshy islands in the gulf at the north of Adriatic. The new city that soon grew from this new source of wealth, Venice, was soon to be enriched by its increasingly dominant role in the European spice trade, a dominance maintained for many centuries (Figure 2.3). By AD 813, Venice claimed a monopoly on all Byzantium trade into Europe, forming an alliance with Constantinople. It was about this time that records of the purchase of spices and documents telling of spices being used as part of the payment of salary become more common. Spices, in their unground form store well, have a high value per unit weight and easily traded throughout the world, hence were an acceptable form of 'currency' (the term peppercorn rent which now means a nominal sum had a different meaning during that period). This currency might also have had an attraction in the ninth century when the price of spices rose very rapidly.

The competition between the northern spices routes leading into Europe via Byzantium and southern spice routes from the Red Sea continued. Venice and Genoa

Figure 2.3. The City of Venice, like some other cities in Northern Italy, was greatly enriched by its participation in the spice trade with the resulting blossoming of the arts, science and architecture.

were to compete for their share of the spice trade into Europe for many decades. Some European traders were happy to trade with the Muslim traders offering spices that were being imported via the Red Sea ports. A visitor to Cairo in AD 996 recorded that the city was host to 160 merchants from Amalfi alone. The cosmopolitan nature of some Middle Eastern cities was remarkable. In these modern times, it is sometimes forgotten that Muslims, Christians and Jews coexisted peacefully in many cities for long periods.[2,3] At this time, the Karimis, a group of Jewish merchants based in Cairo, had a network of trading agents that stretched from China in the east to Mali in West Africa. However, this coexistence and mutual benefit was to be upset at the end of the next century.

The Crusaders, the spice trade and the rise of the Ottoman Empire

In 1095, the first Christian crusade against Islam was launched. There were many motives for participation in the crusades, not all religious. Although Pope Urban preached the radical doctrine of fighting a holy war, some were motivated more by mammon. As the first Crusaders gathered in Byzantium to launch their assault south, some Venetian merchants were well aware of the fact that their traditional spice trade routes were already under threat from Islamic military and commercial competition, so aiding an army that might counter those threats might be a good investment. Some other wealthy Italian trading cities also gave assistance to the Crusaders because the merchants in those cities were given guarantees of important trade concessions in the eastern Mediterranean ports should the Crusaders be successful. These merchants saw an opportunity to compete more successfully with the Venetians by taking control of alternative spice supply chains, currently controlled by Islamic rulers.[4] The first assault failed when the shambolic, undisciplined Christians were routed by a Turkish army. In 1097, more organised, reinforced and professional Christian armies moved again and began to push the Turks south and eventually the Europeans occupied parts of Syria and Palestine. So began a struggle that lasted for hundreds of years as the invaders and local rulers fought over this area, an area so important in the spice trade. The new Christian states in the Levant gave the Europeans improved access to the southerly spice routes and the eastern end of the Mediterranean was safe for the growing Italian fleets which enriched Genoa, Pisa and Venice. The trade in spices increased, prices fell, and the spice market in Europe expanded again. By the twelfth century, the spice trade in Europe had grown to the extent that guilds of spicers and pepperers were formed. By the thirteenth century, specialist spice shops were doing business in some European cities. Although the short-lived conquests in the Levant reduced the power of the Arab traders for a period, inevitably the traders to the east increased their supply via the more southerly Egypt routes, where Cairo and Alexandria blossomed with new wealth.

An important change in the trading relationship between spice producers and spice consumers was about to occur with the growth of the Ottoman Empire (1299–1922). Gradually, the European powers lost control of the Levant and power was also slipping away from Byzantium in Turkey. Osman I started a dynasty which grew over the centuries to cover all the major spice trade routes leading to Europe. By 1453, when the Ottomans took Byzantium, this great Islamic Empire had complete control of the trade between India and the Mediterranean and they profited greatly by this monopoly. The resulting wealth that accrued to the Ottoman Empire from its monopoly of trade (not only spices but silks) between the East and Europe led to a blossoming of Islamic art, architecture, science and theology, such that its culture seemed very sophisticated compared to contemporary European culture (Figure 2.4). The Muslim rulers tolerated non-Islamic, especially Jewish, traders in their midst because these traders had access to the heart of Europe, which assisted their trade. As noted earlier, Italian traders were well represented in Egypt; hence, even when there were tensions between Christians and Muslims, some Christian traders, and in particular Jewish traders, continued to do business with Arab traders. The Venetians in particular were ready to put trade before religious zeal and it took the threat of excommunication of many Venetian merchants by the Pope in 1322 to temper this trade and then only for a short time.[2] Surprisingly, there was less tolerance to Christian traders because of the memory of the terrible acts of violence conducted by the Crusaders in the previous centuries. One wonders whether this Muslim preference for Jewish NP traders might have been one of the causes of Christian resentment of Jewish businessmen so evident in European literature (e.g., *The Merchant of Venice*) and society in general.

Figure 2.4. During the period when the Ottoman Empire controlled the spice routes leading to Europe, it was noted for its flourishing arts, science, medicine and sophisticated culture.

The rise of western European power

The increasing Ottoman control of the spice trade into Europe was to have profound consequences there. Each link of the chain of traders and merchants, stretching from Indonesia to Europe, took their profit as spices passed through their hands and the rulers extracted their taxes. It was not surprising that a 1000% rise in value between the source and the final market place was common. Predictably, this huge economic activity caught the imagination of some merchants furthest from the source of supply of the spices. These merchants wondered if, and how, they could get their goods without paying so much to middlemen. The merchants in northern Europe and those at the west end of the Mediterranean had most to gain by finding a route to the sources of the spices (and silks), a route which would bypass the Muslims in Middle East and traders in the Italian states. The Spanish and Portuguese were the first to turn speculative discussions about the possibility of sailing west to the Spice Islands into reality.[5] The Portuguese had already explored extensively down the east coast of Africa and some navigators were confident that they could find a passage to the east around that land mass. In 1492, Columbus set off to find the western route, with the source of spices being his main goal. He was quite sure he had found his goal when he reached the Caribbean; indeed he collected aromatic plant specimens to prove that he had reached the land of spices. On his return to Spain, he blamed poor storage and a necessity to harvest at the wrong time as the reasons why his 'spice' samples were nothing like the genuine articles. Columbus never fully accepted that he had not found the Indies, indeed parts of the region he brought to the attention of Europeans became known as the West Indies.

Some of its inhabitants were even given the generic name of Indians! Although this Spanish quest for spice led to the discovery of a very valuable source of gold and silver, and subsequently led to European colonisation of the newly discovered continent, it failed in its primary objective. But, as outlined below, centuries later some other important exploitable NPs (found in coca, quinine, capsicum, vanilla, tobacco and chocolate) were to come from this continent.

In 1497, the Portuguese launched their first maritime expedition to the Spice Islands. Three small ships, commanded by Vasco da Gama, set sail and headed south in what was to be a two year ocean voyage covering 24,000 miles. This was to be an epic voyage, surpassing Columbus's feats in both navigation and endurance. Unlike the earlier Spanish expedition, it was to succeed in re-establishing direct European contact with India spice ports for the first time since the Romans. Having made landfall, Vasco da Gama, en route for the city of Calicut, was impressed by the cosmopolitan nature of the area but sadly took little time to consider why this peaceful, civilised society had been so successful and why its ruler, the Zamorin, was so tolerant to so many races of foreign traders. The arrogance and religious intolerance of Vasco da Gama upset the Zamorin, who was also unimpressed by the gifts Vasco da Gama could offer—Europe had nothing very novel or useful to offer the East at that time. Gifts from the Portuguese that had impressed the rulers along the African coast in previous explorations seemed paltry to the Zamorin who regularly hosted richer, more generous traders than the new Europeans. However, Vasco da Gama was given the right to trade and returned home with a load of spices. This voyage was the beginning of the rise to world status of the western European powers. In 1500, the Portuguese dispatched 13 ships commanded by Pedro Cabral, crewed by one thousands sailors, back to India (coincidentally discovering Brazil en route (hence laying the foundation of the Portuguese empire in South America). The intention of this venture was not only to import spices but also to export Portuguese power and influence. On arrival in Calicut, Cabral demanded that the Zamorin expel all Muslim merchants, contrary to his host's long tradition of free trade. The tension, first provoked by Vasco da Gama, now exploded into violence. The Portuguese knew that there was no competition for their naval artillery and they were determined to eliminate all local Muslim competition. They did so in a bloody and terrible way and took control of Calicut. This was the first of many such uses of Portuguese power in the Indian Ocean, an Ocean that the Portuguese came to dominate for a century. The Portuguese were building an Asian trading empire that was to last for 500 years. Unlike the Asian colonies of other European nations, the Portuguese trading empire was based on small, strategic, well-fortified trading posts and not on large territory possessions. By 1501, the Portuguese merchants could easily undercut the Italian spice merchants in European markets. Not only had the Portuguese their own cheaper source of supply but their disruption of the trade on the Malabar Coast had caused a rise in the price of spices imported via other traditional routes. Not content with their wealth, the Portuguese now pushed their way further east along the supply chain. By 1505, the Portuguese were trading cinnamon directly with suppliers in Sri Lanka; in 1511 they had seized control of Malacca, the key port in controlling trade between India and the Spice Islands. In that year, the Portuguese launched an expedition from newly conquered Malacca to find the Spice Islands which they finally located the following year after some epic adventures.[5]

The early western European success in dominating so much of the global spice trade so rapidly fed a hubris that was wonderfully illustrated by the fact that the Pope decided that the world trade would be shared by only two Christian nations! The Spanish were granted exclusive rights to trade in products from one half of the globe, while the other half of the globe was assigned to Portugal by the pontiff. Spain decided that they had right to the Spice Islands, not the Portuguese. This was remarkable given that there were no accurate maps of the world, the earth's circumference was unknown and the Pacific was unexplored by Europeans. However, this dispute was sufficient for the Spanish to seek their own route to 'their' islands. The Spanish decided to finance an exploration by (ironically) another Portuguese explorer Magellan, who proposed to the Spanish that a route around the southern tip of South America would be quicker and more convenient than the route around the Horn of Africa. Little did he know that of the five ships that set sail in 1519 with 270 men, only one ship was to return to Spain and only 18 men survived. This voyage was the first circumnavigation of the globe but Magellan had died in the Philippines long before the fleet reached the Spice Islands. With this visit to the Spice Islands to bolster their claim, Spanish diplomats restated their 'rights' but after further arguments with the Portuguese, the Spanish king, in need of money for his wedding, traded the trading rights to the islands to the Portuguese. This might have been a rational settlement because the Portuguese were much more able to defend their established posts in the east than the Spanish were able to attack them. So, Portuguese hegemony continued.

However, the hopes that the Portuguese might have of establishing a monopoly were doomed to failure because market forces simply stimulated competitors to adjust to the new realities. Muslim traders soon developed such skills at smuggling spices from the many small ports on the Malabar Coast that trade via Alexandria and Venice recovered within decades. The smuggling of NPs is a theme that will reoccur repeatedly and it continues to the present day with narcotic smuggling.

Meanwhile, the Protestant Dutch and English were no more happy with the Catholic dominance of the spice trade than they were with the previous Muslim dominance. English and Dutch 'Privateers' (an English euphemism for what they would call a piracy in others) were gaining wealth, and seafaring experience, by raiding Spanish and Portuguese trading ships returning from their voyages. The Dutch and the English were nations with growing maritime confidence. Both nations had penetrated the Indian Ocean and the Pacific Ocean and eventually their ships turned up in the Spice Islands. Sir Francis Drake had visited the Spice Islands in 1578 and left with an agreement from one sultan that the British could have exclusive rights to the cloves that he could supply. On Drake's return to London it was hardly surprising that some merchants decided that an investment in an expedition to the Spice Islands might enrich them. But with no single merchant confident enough or rich enough to bankroll the expedition, the East India Company was formed in 1600 to spread the risk and the investment. (This model of investment was copied elsewhere: the Dutch East India Company was formed in 1602, the Danish East India Company in 1616, the Portuguese in 1628, the French East India Company in 1664, and the Swedish East India Company in 1731). The British East India Company was to lead British colonial expansion and to develop a remarkable skill at trading many different NPs. The Dutch arrived in the Spice Islands in 1599 and the English in 1601. Neither the Dutch nor the British, being Protestants, took any notice of the 'ownership' of the world as decreed by the Pope and both nations laid claim to any territory they could gain control over by use of force. The Dutch went east, ready to fight for possession, and quickly took over power from the Portuguese, sometimes aided by alliances with disgruntled locals. Sri Lanka, with its source of cinnamon, fell under Dutch control in the 1630s, Malacca in 1641 and the Malabar ports in the early 1660s. Just as the Catholic neighbours from southern Europe had found themselves in conflict over their claims to the Spice Islands and strategic territories en route, the Protestant English and Dutch neighbours now found themselves competing for the same distant parts of the world. The English never seriously challenged the Dutch in the spice trade despite considerable effort. By 1662, King Charles II issued a decree forbidding English merchants to buy cinnamon, cloves, nutmeg or mace from any other than the growers. But this was not an early enlightened example of Fair Trade, rather it was because the English king hated to see his tax collectors in competition with Dutch spice merchants. The conflicts between the Dutch and the English led to war in 1665–7, but a peace treaty gave the Dutch sovereignty over the Spice Islands (which the English could not practically challenge in any case) in exchange for the Dutch giving up their claim to New Amsterdam which the English actually already occupied and called New York. The Dutch economy and culture blossomed in the way that the Ottoman and Venetian economies and cultures had done previously when the spice trading had enriched them. The period from 1620 to 1670 was the golden age of Dutch painting, the city of Amsterdam became one of the richest cities in the world at that time and science and learning flourished (Figure 2.5).

The Dutch East India Company (VCO) ruthlessly and efficiently operated a monopoly. For example, in 1735, the VCO burned half a million kg of nutmeg in Amsterdam in an attempt to raise prices. In Indonesia, the Dutch would torch plantations if new growers tried to enter the market or if established growers tried to smuggle their products to non-Dutch traders.

The remarkable legacy of the spice trade

The fact that the Dutch had to manipulate the market by destroying their once valuable spices tells us that the market that had expanded so dramatically for hundreds of years was becoming saturated—a modern economist would say that the spice market was mature. Consumers only needed a certain amount of spice. Furthermore,

Figure 2.5. When the Dutch gained a near monopoly on the spice trade into northern Europe, the resulting wealth made Amsterdam one of the richest cities in the world, with a consequential blossoming of the arts, education, medicine and science.

other NPs were beginning to interest the palate of the wealthy, with coffee and tea also possessing interesting scents and flavours to tickle the taste buds of discerning consumers. Other economic factors were also at play. As other European nations increased their territorial possessions around the globe, some inevitably gained land with climates that were not unlike the places where spices had been grown exclusively until the seventeenth century. Both the Portuguese and the Spanish had tried to establish spice plantations in their new colonies in South America but without significant economic success. The Dutch were well aware of this potential competition. For example, the Dutch treated nutmeg with lime before export to ensure the nutmeg would not germinate. However, as in the case of all plants with economic importance, some plants will inevitably be smuggled out if the price is right. Peter Poivre eventually succeeded, after many remarkable adventures, in taking nutmeg plants to the French colony of Mauritius. Even though these plants failed to thrive, Poivre's writing about his exploits inspired others. He raised French money and returning to the Spice Islands he managed to smuggle 2000 nutmeg seedlings and 300 clove seeds from the islands. In Mauritius, a few plants grew and they cropped in the late 1770s. However, the new Mauritius producers could not compete with the Dutch, who were still supplying several thousands of kg of cloves every year to France some decades later. But the principle of moving NP-rich plants to new parts of the globe to break a monopoly was proven.

The Dutch lost the control of Sri Lanka to the British and its cinnamon groves in 1795 as the British East India Company expanded and consolidated its power base around India. During the Napoleonic Wars, the British gained control of the Moluccas for some years and they transplanted many of the spice crops to their possessions in Penang (Malaysia) and Singapore. In 1818, the French found that cloves that merely grew in Mauritius thrived in Madagascar and Zanzibar to the extent that two centuries later these islands supplied cloves to Indonesia. In 1843, the English established successful nutmeg plantations in Grenada. Thus by the nineteenth century, the Dutch spice hegemony was broken and for the first time in thousands of years it was no longer possible for states, cities or individuals to make fortunes from the spice trade.

The legacy of all the groups who gained wealth and power from spice trading is still very evident. The current geopolitical map of the world was shaped in part by the Greek, Roman, western European, Muslim and Indian involvement in satisfying the human craving for spices. Languages, cultures and religions were spread as traders sought profitable sources of spices. For example, it was not the British state that imposed the English language around the globe, it was the British commercial interests in NP trading, starting with spices. British power around the globe was initially simply a means of protecting these British traders and much of that power was privately financed and managed. For example, in India it was the East India Company that provided the military power. That power was subsequently (in the second half of the nineteenth century) brought under crown control, and the concept of the British Empire was a late Victorian construction. However, as history shows, the power that accrues to those who have a monopoly on spice trading is easily lost once spices became more like commodity products. The British and Dutch were, however, about to pull off a coup by finding new NPs to excite European consumers.

Beyond spices—NPs shape society, fashion, business and politics in Europe

As the European spice traders encountered novel cultures, they learned about new plants rich in NPs, and these novel NPs soon fascinated them. By encouraging consumers in Europe to share this fascination, there was money to be made. Many of the features of what is now termed 'the global economy' can be traced to this period. In nearly all cases, the initial rarity of the NP-rich product ensured that only a few rich people enjoyed the exclusivity of the product. Given that the rich often liked to show that they had time and money, they usually developed little rituals around their conspicuous consumption of these expensive NPs. The aspiring middle classes, a growing group with most developing European countries, were usually keen to adopt some social behaviour of the upper classes and NP consumption was one activity that was within their reach as the prices of NPs fell due to a rising supply in a competitive market. The middle class might not be able to afford the capital to buy a big house, beautiful horses, land or fine furniture but they could stretch to a cup of tea or coffee and some chocolate cake, as served in the finest houses. The NP traders found that selling their products more cheaply to a bigger market was a natural way of developing their business. Likewise, by encouraging a passing fashion for a new taste, the market for NPs could be kept fresh by introducing new NP-based products at intervals.

The seventeenth century—the fashion of tea drinking

The drinking of herbal infusions had long been known in Europe, mainly in a medicinal context. However, in parts of China and Japan the social drinking of infusions of the leaves of tea plants had a very long tradition, stretching back thousands of years. The Europeans, who first encountered tea drinking, sampled the drink, and being businessmen they wondered if there was a market for the product back home. The only significant producer of tea at that time was China, and the first tea was brought to Lisbon by the Portuguese in the last few decades of the sixteenth century. It was nearly a century, in 1652, before tea was recorded in London and in that same year coffee and cocoa also arrived in England. By the last half of the seventeenth century, there was a major trade in tea between China and Europe and that supplemented the spice trade. This is an example of the trade in one commodity, valued for its NP content, causing the development of another market for a different NP. Not only tea but also 'china' was introduced to European society. This fine, strong but delicate porcelain has taken its name, China, from its country of origin but its introduction to Europe was partly coincidental. The sailing ships that went to China to carry the tea to Europe, like all sailing ships, needed to carry ballast in their lowest holds to establish the correct weight distribution for safe and efficient sailing. Tea itself is light because it is simply dried leaves and the tea had to be kept dry, hence was stored in the more airy upper holds. The ideal material for ballast is something that can itself be traded and porcelain pottery was so cheap in China that there was little to loose if there was little market for it in Europe. In Europe, the pottery being manufactured was usually a heavy earthenware; hence, the wonderful decorative fine china was seen as very attractive, especially by the rich serving the newly fashionable tea. During the period of intensive porcelain imports from China (1684–1791), 24,000 tonnes of porcelain were imported to the United Kingdom with over 200 millions individual pieces being sold. The quality of the cheaply imported fine china was such that European potteries were forced to innovate in order to compete. Eventually, in 1709 the art of making porcelain was developed in Europe with the introduction of Meissen china and later in 1742, similar processes were devised in the United Kingdom using 'china clay' from Cornwall.

The limited competition in tea trading, due to monopolies being granted, usually to the East India companies in each country, allowed governments to tax these imports efficiently. However, this heavy tax inevitably encouraged tea smuggling. By 1770, the 12,000 tonnes of legally traded tea entering England might have been joined by 6000 tonnes of smuggled tea. Scotland and the south east of England were the most active regions for tea smuggling with well organised and violent gangs providing an early example of the way in which crime thrives when governments attempt to control NP supply (see sections on tobacco, opium, cocaine and cannabis). It has been argued that tea smuggling had a significant impact on the profits of the East India Company; indeed, the company was in serious financial difficulties in the 1770s. The company successfully petitioned the UK government for a monopoly on tea supply to North America and the company paid more attention to the emerging North American market where smuggling was less damaging to its interests. However, the English colonists in America resented this monopoly of tea supply, and the tea tax, so they added this grievance to others (tobacco taxes were possibly even more irksome) and eventually fought for their independence from Britain. When the tea tax in the United Kingdom was reduced by a factor of 10 in 1784, tea smuggling became unprofitable and tea consumption increased greatly. The percentage of the UK population that could afford tea increased and tea drinking became part of the British culture at all social levels. The removal of the East India Company UK monopoly in 1834 further reduced the price of tea but by then the company had found another NP-rich monopoly to exploit—opium. One problem that the European traders faced when trading with the Chinese for tea (and to a lesser extent silk) was that the Chinese merchants wanted rather few goods or products that the Europeans could offer. The Chinese (like their neighbours, the Japanese) were remarkably self-sufficient. The English East India Company could offer the Chinese copper but the company still had to part with large amounts of silver to purchase the tea they needed. Where was that money to come from? The answer was another NP, the morphine contained in opium, which considered in detail later. However, the East India Company had another strategy—to break the Chinese monopoly on tea production. It first planted tea in Assam in 1835, and sold its first Indian tea in London in 1838. In the 1840s and 1850s, tea cultivation spread to Ceylon. Darjeeling joined the tea growing areas in 1856. In the last half of the nineteenth century, China tea lost its popularity in Europe. China tea held over 90% of the market in 1866 but less than 10% by the end of the century. Tea had become a commodity product and in the twentieth century several nations were competing in the market (India, China, Sri Lanka, Indonesia and nine nations in East Africa, led by Kenya). The breaking of the monopoly on the supply of tea had been accompanied by a breaking of the monopoly on the supply of caffeine in tea. The total world production of tea is around 3 million tonnes per annum, worth $9 billion to the producers.

Coffee—the Ottoman's favourite stimulant becomes the world's most important NP-rich drink

The drinking of tea had provided the main stimulant (caffeine) in the Far East for over 4000 years. Another source of caffeine, coffee, has a much shorter, but equally interesting and important history.

Coffea arabica is a native of Ethiopia. It was grown in some regions bordering the Red Sea and was traded throughout the Arab world. The chewing of the beans was a practice adopted by those seeking an aid for keeping alert on long journeys or, it is said, attending long prayer meetings. Coffee consumption has been known in the Middle East and further east from about 800 BC. Later Ottoman scholars traced the history of coffee drinking in their empire using Islamic edicts which reveal discussions of the properties of the drink and whether its consumption was consistent with Islamic law. The consumption of alcoholic drinks was banned by Islamic law, so the fact that coffee had some physiological effects concerned some clerics. The debate was won by those who liked its mild stimulant effect, a majority that included many influential rulers. It was the Turks who developed ways of roasting and processing the beans and using them to make a drink.[6] Coffee reached the capital city of the Ottoman Empire, Constantinople, in 1453 and coffee shops became increasingly important places for social discourse (Figure 2.6). Coffee was being drunk by Sufis in Cairo by the early sixteenth century and in Aleppo (an important spice trading port in northern Syria). The first documented European record of coffee was in 1573 when Leonard Rauwolf provided the first botanical description of *C. arabica* seeds. Through Rauwolf's writing and his collected specimens, knowledge of the plant, and the drink derived from its roasted berries entered European scientific discourse.[7]

It was through links with spice traders that coffee was introduced to the United Kingdom (and some other nations). The Company of Merchants of England Trading into the Levant Sea (incorporated in 1605) was a very rich group of UK merchants who were given a monopoly of trading between England and the Ottoman Empire. Spice trading was a significant part of their business. Some of these merchants lived in large, very affluent European enclaves in Syria, along with Greek, Italian, Dutch and French traders. In 1651, one of the traders in Smyrna was Danial Edwards who came back to

Figure 2.6. The coffee shop, as a place to enjoy social interactions and to benefit from business deals (at least for males), began in the Ottoman Empire and was adopted throughout the rest of the world. It is a rare high street in any city that does not now have at least one coffee shop.

London bringing with him his Greek assistant Pasqua Rosee, some coffee beans and the apparatus for roasting the beans. London at that moment was under Puritan influence with alcoholic drinks being frowned upon; hence the drinking of a new stimulant appealed to Edwards and his friends. The popularity of the social coffee drinking at the Edwards house gave Edwards and his father-in-law Thomas Hodges the idea of opening a coffee shop. In 1654, they opened a stall, managed by Pasqua Rosee, on a site near the Royal Exchange, an area frequented by merchants in London. With a thriving business, Rosee was joined by Christopher Bowman and moved the business in 1658 to new premises, but Rosee encountered increased prejudice as a foreigner so he left England to start a coffee shop in the Dutch capital, The Hague. The success of this first English coffee house encouraged others to start similar coffee houses and Europe's first coffee culture was flourishing by 1660. Not only did merchants favour these new coffee houses, but writers, philosophers and scientists also found the company and atmosphere stimulating.[7] Coffee houses became the places where news was exchanged, politics discussed (this was a period of intense political debate in England as the roles of the parliament and the crown were under scrutiny) and business transacted. As the number of London's coffee houses increased (in 1739, a survey counted 551 in a city of 500,000), some were favoured by particular groups and one can trace the founding of some important commercial institutions to individual coffee houses. In Garraway's, the first lists of share prices was maintained and other coffee houses kept list of commodity prices, currency exchange rates and the price of government bonds. Mr Bridges Coffee House maintained a list of goods recorded by the Custom's Office. As early as 1692, Edward Lloyd published his shipping list at his coffee house and those seeking shipping news increasingly congregated at Lloyds. This collection of coffee drinking merchants, brokers and underwriters involved in shipping organised themselves into a group, left the old coffee house, founded their own more exclusive coffee house, which became Lloyds of London. Not until 1773 did Lloyds of London leave behind the making of coffee to concentrate on the making of money.

In 1761, at Jonathan's Coffee House, a group of stock jobbers (share traders) who met there regularly to conduct their business (aided by paper, quills and ink provided by the coffee shop owner) concluded that their affairs could be conducted more pleasantly if they had sole use of the coffee house. They negotiated a 3 hour exclusive daily use, but when challenged in the courts by someone excluded from the group they decided to leave Jonathan's and start their own club in Threadneedle Street. The New Jonathan's included a large coffee room but its name was soon changed and it became the London Stock Exchange (The New York and Boston Stock Exchanges also grew out of their coffee house cultures).

In many other European cities by 1700 coffee houses began to appear, attracting the wealthy and fashionable. As in London, gentlemen, scholars and merchants deemed the new coffee houses to be the place to meet like-minded people. Meeting for coffee with the real intention of meeting to socialise, gossip, do business and to plot was now part of the European culture. As the demand for coffee grew, coffee merchants began to worry about the near monopoly of supply by the Ottoman Empire—the experience of the spice trade monopolies were still fresh in the minds of many merchants. Inevitably, thoughts turned to finding an alternative source of supply. For centuries, the Ottoman traders had very effectively ensured that coffee plants and viable seeds were kept away from those who were buying one of their most valuable products. The Europeans had broken the Arabian spice monopoly, so the coffee monopoly was worth guarding. However, a few coffee plants had already been grown in India and in 1696 the Dutch East India Company obtained some of these plants and shipped them to their colony in Indonesia. By 1720, these new plantations were producing significant amounts of coffee for export to Europe. Coffee plants were also sent to Amsterdam, where they were grown in the Botanical garden, and these plants provided a source material for the French to take to their Caribbean colony of Martinique. By 1712, the Dutch had coffee growing in their South American colony Surinam, and coffee reached the Portuguese Brazilian colony in 1729 and later Columbia. The British established plantations in Jamaica in 1730 and later introduced the crop to East Africa. So by the eighteenth century, the 'colonial' traders (the same merchants who had been trading spices and tea—predominantly the Portuguese, Dutch and the British) had successfully taken coffee to other continents. The Ottoman monopoly was broken and several nations competed to supply coffee to the growing markets in Europe and North America.

Coffee shops remained an important part of European culture in the nineteenth century, but coffee consumption at work and in home increased throughout the industrial world. So embedded is the consumption of coffee in business, academia and social affairs that a day without access to coffee is unusual. Business meetings often require the supply of coffee. Scientific meetings always have coffee breaks (which many think are the most productive parts of the meeting) and many scientific institutions have 'coffee clubs' where research is discussed in a less formal way. In social life, 'drop by for coffee sometime' or 'would you like to come up for a coffee' had all sorts of interesting hidden meanings. Although the monopoly of coffee supply was broken in the eighteenth century, a new form of monopoly was developed in nineteenth and twentieth centuries that enabled huge amounts of money to be made from coffee again—the coffee brand. Anyone can buy coffee beans quite cheaply but only Starbucks can sell Starbucks. By clever marketing, a few companies buy coffee beans very cheaply and using the monopoly of a brand they now control the mass market for coffee. The way in which consumers award a near monopoly to a brand, despite perfectly good alternative sources of supply, changes the rules of the NP economic model and we shall find more examples of this remarkable consumer behaviour later.

The annual world coffee production is about 8 million tonnes, worth about $16 billion to producers. Coffee is the second most valuable legally traded commodity after oil and 25 million people worldwide gain their living from coffee. Starbucks purchase about 140,000 tonnes of coffee each year and customers spend about $4 billion in their cafes; coffee is the largest imported food in the United States. A remarkable 2.25 billion cups of coffee are consumed every day—20% in the United States.

The South and Central American chemical gems—coca, quinine, tobacco and cocoa

The Portuguese were displaced from the East Indies trade before the tea and coffee trades expanded and the Spanish never managed to gain exclusive access to any Far East products so they have featured rather little in the NP products story so far. However, these southern European nations were to find several useful NPs in their South and Central American colonies.

Cocoa—Central America's great contribution to human happiness

The NPs found in cocoa were to bring great pleasure to many throughout the world in the form of chocolate. The key NP in chocolate is the bromine which is a pleasant stimulant related to caffeine. Although valued as a drink in Central America for centuries, the first shipment of cocoa beans only reached Spain in 1585. The drinking of chocolate was taken up by the Spanish monarchy and spread to the French court when there was a marriage between the two families of ruling monarchs. The French King Louis XIV was especially fond of chocolate. By the mid-seventeenth century, chocolate had become fashionable among the rich in Paris, helped by its reputation as an aphrodisiac. In 1657, the drinking of chocolate was introduced to London and quickly became fashionable. The fashion spread to many European capitals during the last half of the seventeenth century and cooking with chocolate became increasingly common in rich households. By the start of the eighteenth century, there were signs of a wider use, as evidenced by a tax imposed on it by Germany. By the start of the nineteenth century, chocolate was being processed on an industrial scale and during that century the market grew as prices fell and the number of consumers grew. The development of the chocolate press in 1828 allowed the cocoa butter to be separated from the remainder of the bean so producing a better, cheaper drink and also allowing the development of what we would now call chocolates. Some of the major multinational industrial chocolate manufacturers can trace their roots to this period. In 1857, the cocoa plant was being cultivated in West Africa with Ghana being especially successful at producing the crop.

The fact that drinking chocolate was regarded as a healthy and acceptable drink, even by those who denounced other stimulants, encouraged the Quakers to become involved in its manufacture. The great Quaker chocolate industrialist in England, the Cadburys, the Frys, the Rowntrees and the Terry's not only made their products widely available but also they used some of their NP-linked wealth to encourage serious social reform. Indeed the United Kingdom's most important and influential source of social research is still the Rowntree Trust, so society can once again thank the human obsession with NPs for a social good.

Coca

Fifteen of the over 200 species of Erythroxylon produce coca but E. coca and E novogranatense now dominate coca production. These two small bushy shrubs grow well in the

Andean rain forests (Columbia, Bolivia, Peru and Ecuador). Coca thrives well in partial shade, hence can be grown among other forest trees or it can be grown in plantations. The leaves of the tree were widely used by Andean residents for two millennia before Europeans first encountered the practice.[8,9] Chewing the coca leaves gives the consumer, whether human or animal, a sense of well-being, competence and energy. Tasks are easier to perform and to sustain. Needless to say, such attractive attributes came to the attention of the military and in the 1880s the Bavarian army obtained cocaine from Merck to try on soldiers. The troops showed greater endurance, were more cheerful in the face of adversity and most importantly required less food and drink. The most attractive of these effects to the military was the loss of appetite because it was estimated that a valuable 15–20% reduction in the transport of provisions could be gained. Some members of the medical profession were as impressed as the military. Cocaine was found to be a good palliative for toothache and it was an effective local anaesthetic. Sigmund Freud experimented with the drug himself for 3 years and also prescribed it for some of his patients (and supplied it to fellow doctors and even some students in the Medical Faculty of Vienna University). Freud found that the drug helped his depression, fatigue and insomnia. He argued that cocaine was not addictive but other scientists were starting to report negative effects that included addiction and severe psychological effects in some who partook of the drug. However, there were no restrictions on the use of cocaine (or indeed most drugs) in the nineteenth century and the use of cocaine, or less pure coca extracts, increased. Because cocaine has an ability to reduce mucus secretion and swellings, it was a common ingredient of medicine sold to alleviate the symptoms of catarrh and malaria. Some medicines contained mixtures of cocaine or coca extracts, phenacetin (an early synthetic medicine discovered in 1887) and another NP quinine. However, the most widespread use of the chemicals in the coca leaf in western society between 1850 and 1914 was in 'patent medicines' that were available from doctors, pharmacists and even the corner shops. All manner of imaginative mixtures, differing in colour, consistency, taste and smell were sold as 'pick-me-ups'. Coca extracts were eventually to find occasional uses in a chewing gum for those with toothache, cocktails and even ice cream and fruit cordials.

One of these cocoa-containing mixtures was to gain a particularly fine reputation and was to have a profound effect in influencing western culture. Angelo Mariani, a Corsican, devised a tonic by steeping coca leaves in red wine for 6 months. This tonic wine gained the endorsement of three popes, six heads of state (including Queen Victoria and US Presidents Grant and McKinley), Bleriot, Ibsen, Rodin and HG Wells. The recipe for this product was so successful that there were many imitators. One of these was John Pemberton, a pharmacist from Atlanta, who like many such pharmacists made up mixtures containing many attractive NPs—in his case it seems he used caffeine, coca leaf extract, cinnamon, nutmeg and vanilla in a base of sugar, phosphoric acid and alcohol. However, Atlanta banned sales of alcohol; hence, Pemberton devised a non-alcoholic syrup that could be diluted with water and could also be carbonated. Extract of kola nut was added to the new tonic cordial and Coca-Cola was introduced to the world. It was a failure. Pemberton sold less than 800 litres in 5 years, so he sold his business to

Figure 2.7. The importance of NPs in the manufacture and marketing of soft drinks has grown since western consumers have demanded 'natural' ingredients. A recent advertisement for the drink 'Red Bull Cola' nicely illustrates this fact.

Asa Griggs Chandler, another, more astute pharmacist who saw the potential in Coca-Cola as a soft drink rather than a tonic. In the next 15 years, he made a fortune from the Coca-Cola company and founded what was to become the world's best known multinational. The formula of Coca-Cola has changed over the years and the coca extract was removed in the early twentieth century and many decades later caffeine-free versions became available. Competitors produced their own versions of what became known as cola drinks, nearly always containing caffeine and a variety of natural and synthetic flavouring (Figure 2.7).

Coca and cocaine

It seems that the chewing of coca leaves and the ingestion or injection of the pure alkaloid cocaine produced rather different symptoms. This is not surprising because even if the main physiologically active ingredient is cocaine in both cases, the dose and purity of the ingested chemicals will differ. A chewed leaf is likely to release its chemicals much more slowly into the bloodstream than ingestion or injection of the pure chemical—the dose profile with time will be very different. It is also quite possible that coca leaves release other physiologically active substances that may themselves have physiological effects or may modify the effects of the cocaine. It seems that the Andean cultures, over the two millennia use of coca, had found a way of exploiting the beneficial aspects of coca ingestion without encountering the undoubted negative effects. In contrast, western cultures found little attraction in the use of cocoa leaves when it could 'benefit' from the technology it possessed to obtain the purified active ingredient. The fact that the growing pharmaceutical industry initially saw only the benefits to be gained by prescribing their purified alkaloids (and preferably patented) was to set a pattern which has repeated itself.

The illegal cocaine market

Unfortunately, addiction is such a complicated psycho-physiological process that it is difficult to judge whether modern, western society could benefit from some of the reported positive effects of the coca leaf without encouraging the addiction of many of its citizens. Although cocaine in pure form is more addictive than many other drugs we ingest, western societies have largely learned to live with the milder addictive substances such as alcohol, tobacco and caffeine. However, in all cases, some individuals suffer the negative effects of their addiction while many get the benefit. Because coca is easy to cultivate in several South or Central American countries (and was successfully grown by the Japanese in Taiwan and by the British in India) and because it has been part of the Andean culture for many generations, it is still widely grown despite the attempts of national and international agencies to abolish its cultivation. If history tells us anything, it is that humans crave the effects of certain NPs so much that they will expend great effort and money to obtain them. Hence there will always be a market for plant material containing stimulants. Consequently, as predicted by free market economics, growers will be found to produce the plant, processors will make a living by practising their skills and a sales and distribution chain will develop to connect the component parts of the business. A poor farmer in South America can expect to harvest two tonnes of coca leaves per hectare (with three leaf harvests per year being possible over the 20+ years of the perennial plant). The leaves contain about 1% of impure, mixed alkaloids and simple processing will give 15 kg of pure cocaine. This material fed to the illegal street market in North America could sell for over a $1 million; hence, the monetary yield per hectare is vastly in excess of any other plant material that could be grown by the farmer. Consequently, it is hard to offer sufficient financial incentives to make coca cultivation unprofitable. It has been estimated that the real production costs of pure cocaine could be less than $1000 per kg yet street price in London in 2007 was approximately 100 times more.

Although cocaine became an illegal substance in the United States in 1914, its use continued, especially in 'creative' industries such as the cinema and theatre (some opera singers also found it beneficial) before the Second World War and again in the latter part of the twentieth century in the music, media and financial industries. Such is the potency of pure cocaine that less than 500 tonnes would supply the current US market of an estimated 10 million users. The funds flowing from the United States to the economies of the four main countries that supply the drug (Bolivia, Ecuador, Chile and Columbia) may account for up to 33% of the GNP of these countries. These vast sums suggest that even larger sums would be needed to eliminate the supply from these sources and the fact that the coca plant will grow on other continents would suggest that the cultivation of coca will be very hard to control.

Quinine—vital drug and a pleasant drinks mixer

Quinine is an alkaloid that can be extracted from the bark of the *Cinchona officinalis* (which also contains the related compounds such as cinchonine, quinidine and cinchonidine). The tree, and 70 related species that contain lower concentrations of these substances, is found in the Andean mountains in regions of Peru, Bolivia and Ecuador. These trees grow naturally at elevations between 700 and 300 m on the slopes of valleys about 10–15 degrees either side of the equator. The bark of the tree was used by the indigenous people to treat fevers and the early Europeans learned of its value. In 1638, the wife of the Viceroy (the Countess of Cinchona) was dangerously ill with malaria in present day Lima, Peru. The court physician suggested that the local remedy quinquina be used.[8] The patient recovered and the Europeans named the genus of the tree named after her. On her return to Spain some years later the Countess took with her more of this useful bark and used it to treat fevers of those who worked on the estates of her husband. The value of the bark was soon known to many in South America and Europe and a trade of the bark soon developed. The Jesuits used their organisation to collect large amounts of the bark from Peru, Bolivia and Ecuador and used the money so gained to fund their missionary work. By the end of the seventeenth century, extracts of the Cinchona, made by steeping the bark in white wine, were used by some of the richer members of the society to treat fevers (especially malaria). The need for an effective treatment for malarial fevers stimulated the use of this treatment throughout the eighteenth century and very large numbers of Cinchona trees were destroyed for their barks. The German naturalist, Humboldt, in 1795 estimated that 25,000 trees were being felled annually in the Loxa region alone and was concerned about the sustainability of the trade. The Jesuits were also conscious of the need to harvest the tree sensibly and they tried to ensure that one tree was planted for each one felled. By the time the former Spanish colonies became independent (1810–1830), demand was outstripping supply and the problem of continuity of supply was made worse by the political instability that characterised that period in South America. However, the destructive harvesting in the Andeas continued for some decades until other sources undercut the South American producers.

The value of the quinine containing bark was now very evident in the growing British colonies in India and elsewhere. Not only were British troops and administrators often weakened by fevers but the local labourers were less productive if suffering from malaria. The economic incentive to find alternative sources of supply eventually encouraged the British and the Dutch (with colonies in the East Indies) to take Cinchona trees to their new colonies where they sought conditions similar to those found in the Andean valleys. These attempts were successful and by the middle of the nineteenth century, plantations of Cinchona trees in the Nilgiri Hills in southern India were well established. By experimenting with harvesting regimes (coppicing and selective bark peeling rather than the destructive harvesting of the tree) and cultural conditions, great improvements in both the yield and the quality of the product were made. The harvesting methods also increased the percentage of alkaloid from about 4% to nearly 7%. By 1880, these new plantations were mature and dominating the market, so the production in South America declined rapidly. The British and the Dutch production served different markets. The former used nearly all their production for their own colonies, while the Dutch took over the rest of the world market and operated a cartel for many years. The Dutch plantations in Java supplied 80% of the world market until the Second World War, when they lost their colonies to the Japanese, and western allies lost ready access to the drug that allowed its soldiers and administrators to operate in areas of high rates of malarial infection. Fortunately, the importance of quinine had been a very significant factor in the rapid development of the science of chemistry in the nineteenth century and just in time synthetic quinine substitutes were ready to fill the sudden and very important new demands.

The size of the market for quinine, the growth of that market and the insecurity of supply in the middle decades of the nineteenth century corresponded with the great advances in synthetic and analytical chemistry at that time. By 1834, the German chemist Runge had attempted to synthesise quinine from coal tar but made quinoline instead. However, encouragingly in 1842, the important alkaloids found in Cinchona bark had been shown to degrade when treated with the strong alkali caustic soda to give the compound that Runge had made, quinoline. The next advance was made by an 18-year-old English student, William Perkins, who was trying to make quinine from coal tar when working during the holidays in a home made laboratory. Like Runge, he failed and like Runge his product was to be more important industrially as a dye than a drug. The messy tar-like product in Perkin's reaction flask, when washed out with water, formed a deep purple coloured solution. Perkins immediately realised that the highly pigmented material might be useful as a dye and he tested this by dipping some fabrics in the pigmented solution. The dye was found to permanently dye wool and the purple colour, mauve, quickly became the fashionable colour of society. Perkins established dye works, made his fortune within 20 years then sold his company to a German competitor and retired to the life of an amateur chemist and patron of other scientists.[10,11]

Quinine—a popular tipple

Low concentrations of quinine in water (about 80 mg/L) have a sharp, bitter taste that some people find attractive. The first known such drink was lemonade, sold in New Orleans in 1843. The best known commercial product, Schweppes Indian Tonic Water, a carbonated, sweetened quinine solution is often used to dilute the juniper flavoured (due to NPs) gin to produce the very popular drink Gin and Tonic.

There are some[8] who have argued that without the use of Cinchona extracts, and later synthetic quinine analogues, world history over the past two centuries would have taken a very different course. It is argued that the use of quinine to treat malaria (and to a lesser extent yellow fever) facilitated colonialisation by western European countries of territories that would have been too hostile for foreigners with little natural resistance to parasitic-induced fevers. For example, the Panama Canal might not have been built without access to quinine. The outcome of the American Civil War has even been speculated upon because the successful blockade of the Confederate ports caused severe shortages of quinine so the Confederate Armies were debilitated by fevers.

Tobacco—the first industrial exploitation of a harmful NP

> *A lone man's companion, a bachelor's friend, a hungry man's food, a sad man's cordial, a wakeful man's sleep, and a chilly man's fire.*
>
> *—Charles Kingsley, author of The Water Babies and Westward HO!*

There are 66 species of the genus that includes the tobacco plants (*Nicotiana tabacum*). These species are found worldwide and some of these have provided hundreds of millions of people with their daily dose of nicotine for a very long time.[12–14] There is evidence that the stimulant effects of the alkaloids found in several members of this genus were discovered independently by humans in different parts of the world. Although most of the early European explorers to the New World noted the curious local habit of smoking, *N. tabacum* only became of interest in Europe in 1559 when the French ambassador to the Portuguese court, Jean Nicot, sent seeds to the French Queen to be grown as a pretty garden plant. The source of the seeds was the Portuguese colony of Brazil but it seems that the species originated in the Andes and had been taken to the length and breadth of the New World thousands of years ago. In all parts of the New World, tobacco was regarded as a source of medicines but also had specific cultural and ceremonial uses. However, *N. tabacum* did not have a monopoly of human interest. When at the start of the seventeenth century, Europeans first encountered the Amerindians in the English colony of Virginia and they found a culture of smoking *Nicotiana rusticum*. Likewise, over 100 years later, the first English settlers in Australia found some Aborigines using *Nicotiana bethamania*. When the early European explorers penetrated the interior of Africa in the nineteenth century, they found smoking of tobacco to be an important part of social interaction. However, it was *N. tabacum* that was to be distributed throughout the world and processed on an industrial scale. The species was subjected first to selection, then intensive breeding and more recently genetic manipulation. Slowly, *N. tabacum* became the basis of a very major world industry. In some ways, *N. tabacum* was an unlikely plant to be used to gain great wealth because

- it is a remarkably adaptable species—it can be successfully grown on all populated continents, hence it is hard to create a monopoly of supply;
- the processing of the plant is relatively simple—a gardener can produce enough, acceptable tobacco for their own use with little more skill than they need to grow vegetables; and
- it is easy to store, distribute and transport.

So why was it possible for certain countries, some companies and a few people to accumulate great wealth from producing and distributing a NP-rich product that was in theory easily accessible to most consumers? At first it was state power that was used to create and maintain a monopoly of supply with the aim of raising taxes. Several major states are still involved in the nicotine trade, even though they recognise that tobacco consumption can be harmful. It seems that states, as well as individuals, can become addicted to tobacco.[15] However, after state monopolies began to fail, tobacco manufacturers developed a new convenient tobacco variant (the cigarette) that could not be duplicated by amateur growers. A few manufacturers soon recognised that by careful marketing they could make tobacco consumption more widely acceptable in society by advertising the virtues of cigarettes.

One part of the tobacco story illustrates, yet again, the role of NPs in human ceremonies and ritual. The ceremonial use might be at the state level (Amerindian), the tribe level (North American Indians), the home level (the after dinner cigar taken while the women withdraw) or the person level (the individual sitting down to take a cigarette, smoke a pipe or flourishing their snuff box with their own little quirks).

1550–1850—tobacco used in small amounts as a mild stimulant

Throughout this period, the way in which tobacco was used in Europe and North America was confined largely to 50% of the population—men. The way in which tobacco was taken depended on class and nationality.

- Smoking was largely confined to pipe smokers or cigars. In some social situations, a pipe of tobacco was shared by being passed around. Pipes were usually made of clay in this period in Europe but very elaborate wood pipes were used in Amerindian cultures and in Africa, where the communal smoking had a profound social significance; consequently, pipes were works of art. Pipe smoking in Europe was limited by the fact that the cheap clay pipes were weak, easily broken and were hard to light until friction matches were invented in 1827 and slowly improved in later decades. It was also hard to carry out manual work while smoking a fragile clay pipe, so smoking at work was uncommon. Cigar smoking became more common in the middle of the nineteenth century, with Cuban cigars becoming the smoke of choice by the affluent in the United States and Europe. Although there were a few exceptions, it was usually, socially unacceptable for women to smoke pipes or cigars.
- Chewing tobacco was very common in the United States, until the mid-twentieth century and was common elsewhere among manual labourers (e.g., miners could not use naked flames to smoke during their long shifts and seamen found it hard to smoke while at sea). The preparation of the chewing tobacco could sometimes involve the addition of other NPs, such as liquorice to add to the flavour. In the nineteenth century, most tobacco consumed in the United States was chewed. However, the stained teeth, strong mouth odour and need to expectorate at frequent intervals made tobacco chewing increasingly socially unacceptable. By maintaining a 'chaw' of tobacco in the mouth for up to an hour, salivation was stimulated such that up to 250 ml of extra saliva was generated and had to be expelled into spittoons or on the ground. As a visitor to the United States, Charles Dickens was appalled by 'the filthy custom' and was horrified that the 'odious practice' was even accepted by judges, medical students and US senators at their places of work.
- the taking of snuff, the sniffing of powdered tobacco (sometimes scented with other NPs) via the nose, was the most socially acceptable form of tobacco used in Europe

until the development of the cigarette. The flourishing of an ornate snuff box was a sign of wealth and offering the precious contents to a guest was a token of generosity and friendship.[16] A few rich women felt comfortable using a little mild snuff but heavy use caused soiled handkerchiefs which would have been considered unladylike.

Nicotiana tabacum was grown in many parts of the world on a small scale by the seventeenth century but one part of the world was soon to dominate commercial tobacco production—Virginia (and then Maryland and North Carolina). The first settlers in Virginia found the Amerindians growing their tobacco and the settlers realised that they could adapt and develop those proven methods to produce *N. tabacum*. Although the English king James I hated tobacco use (he published his 'Counterblast' to make his opposition clear), he also realised that it would be hard to abolish its use (at that time, about 25% of the UK men smoked tobacco more than three times a week). Instead, James decided to tax tobacco and to give the new Virginian colony a monopoly in supply. This was a clever scheme at that time (although, maybe, not a good long-term move for reasons we shall soon learn) as it gave the new colony a source of income to pay for the import of the manufactured English goods and it was easier to collect the tax on tobacco when it was imported at English ports than it was to collect the tax on any UK tobacco production. Once again a tax on an NP was seen as an excellent way of raising revenue—by 1660 tobacco tax accounted for 25% of all UK tax revenue. The tobacco growers of Virginia expanded production and trade between the English colonies and the home land flourished. However, tobacco is an extremely greedy plant to grow and a good crop could only be produced for one or two years before the soil was exhausted (the removal of most of the plant biomass from the field at harvest inevitably depletes the field of minerals). Although fields could be left fallow to rebuild some fertility, enough 'new' land was available within the state (and later within neighbouring states) to enable tobacco production to be continued and even expanded. This availability of low cost and fresh land gave the North American producers a very real advantage in competition to European production. At that time, the shipping of commodity products from North America to Europe would not have been economic, so a high value NP-containing product was extremely valuable to the new colonies. The tobacco trade expanded throughout the seventeenth century and into the eighteenth century, enriching many tobacco growers who enlarged their estates, built grand houses and developed into a class of well-educated, rich men of influence. This group began to resent the fact that their English rulers demanded not only that tobacco was shipped in English ships but also that tobacco had to be landed, and taxed, in England before re-export to mainland Europe. Inevitably the smuggling of tobacco became increasingly appealing, even to some in the United Kingdom. After the union of Scotland and England, there were no trade barriers between the two former countries, but the new UK government found it hard to enforce collection of taxes in Scotland.[17] Scottish merchants exploited this weakness ruthlessly and flourished. Between 1707 and 1722, the Scots paid only half the duties owed. By 1750, 10 million kg were being landed annually in Glasgow. Scottish merchants became very active in Virginia, buying stock and warehousing the tobacco for export to their home port. By 1760, nearly 50% of tobacco from the Chesapeake area was smuggled and imports to Glasgow exceeded the total of all other UK ports combined.[17]

1860–1960—the rise of the cigarette brands dramatically increases the use of tobacco worldwide

Not all European governments encouraged tobacco consumption to raise taxes; for example, smoking was banned in Berlin streets until 1831. However, those governments enjoying the tobacco taxes would have been thrilled to know that tobacco consumption was about to grow massively with the development of the cigarette making machine.

Cigarettes, as their name suggests, were developed from cigars which came in various sizes. In Seville, small *papelotes* were devised where paper was used instead of a tobacco leaf to form the outer casing of the little cigar and it was realised that leaf fragments could be placed inside these wrappers rather than rolled leaves as used in cigars. They soon became popular with consumers because it was easier to disguise smuggled tobacco inside these *papelotes* and predictably in 1801 the state tried, unsuccessfully, to ban them. The manufacture of papelotes, or cigarettes as they are now known, spread slowly throughout Europe but they remained expensive because each had to be handmade. In France, the state tobacco monopoly began making cigarettes in 1845, and the first British cigarette factory was set up in 1856 by Robert Peacock but was followed soon by the now famous name of Philip Morris. But in Britain and the United States, cigarettes were seen as effeminate so were insignificant in the trade at that time but this attitude was changing. Between 1875 and 1880, annual cigarette consumption in the United States suddenly grew from 42 million to 500 million. No one reason can be given for the change in attitude, but the introduction of the cardboard cigarette packet, brand advertising and the safety match made cigarettes the most convenient and socially acceptable way of accessing nicotine. The brilliant idea of inserting colourful, interesting cigarette cards into cigarette packets appealed to consumers, especially children. So great was the demand for these nicely packaged, well-advertised goods that the manual production became a problem—a skilled woman could make 5 cigarettes per minute and her labour accounted for 90% of the cost of production. In 1880, 21-year-old James Bonsack designed a machine to make over 200 cigarettes per minute, a machine that could work continuously and could produce a superior uniform product. In the United States, the first tobacco manufacturer to realise the potential of the Bonsack machine was Buck Duke, who came to a very favourable agreement with Bonsack and gained an advantage over his competitors that Duke was to exploit ruthlessly. Within 5 years, Duke was selling 2 million cigarettes per day (more than the French sold in a year). Duke spent lavishly on advertising, took over competitors or drove them out of business by undercutting them. By the start of the twentieth century, Duke had gained such a monopoly in the United States that the government forced him to split his American Tobacco Company into three separate companies. Duke left the industry and gave much of his wealth to found Duke University.

In Britain, the Bristol Company of Wills introduced some Bonsack machines in 1883 and was selling 11 million cigarettes per year by 1888. By 1891, they were selling over 85 million cigarettes per year and the business boomed. Wills led the consolidation of the industry in the United Kingdom by negotiating the formation of Imperial Tobacco, which was formed from the 13 leading UK companies including Wills. Duke's attempt to enter the UK market in 1901 was thwarted but only when Imperial Tobacco and Duke's American Tobacco agreed to carve up the world markets between them by forming a joint company British American Tobacco (BAT). The tobacco industry in the United Kingdom had a strong negotiating position because of their dominant presence in the British colonies. For example, there were more consumers in India than in the United States, and India was second only to the United States as a tobacco producer at the end of the nineteenth century. The UK and US tobacco industries were to dominate the worldwide tobacco business throughout the twentieth century.

As cigarette consumption grew in the first few years of the twentieth century, several individuals and groups spoke out strongly about the dangers of smoking but they made little progress until decades later. The First World War saw soldiers being given tobacco as a way of maintaining morale, because a cigarette gave a soldier a few moments of calm before battle or when wounded. A captor might give a captive a cigarette as an act of kindness. By the 1920s, cigarette smoking began to spread to women, helped by film makers showing worldly, beautiful, successful women using cigarettes. By 1940, the per capita cigarette consumption in the United States was 2500 and in war torn Europe packets of cigarettes became an informal currency. But the critics of tobacco use were finally getting a hearing because medical evidence was accumulating that tobacco consumption was harmful. By the time the US Surgeon General declared tobacco use harmful in 1964, over half the US males were smokers but only some gave up their tobaccos. By 1973, the average American (15+) annually smoked over 3800 cigarettes, compared to 3200 in Japan, 3200 in the United Kingdom and Italy and 2700 in West Germany. As smoking was banned in public spaces in the developed world, as health warnings on cigarette packets became more graphic, the big tobacco companies simply turned their attention to the less developed world. The companies see plenty of potential in the 80% of the planet's population who do not smoke currently, especially as in the less developed countries there are thousands of millions of young people who are especially susceptible to advertising. The big tobacco companies boast that they help governments collect tax efficiently (the company BAT claims that in 2006, it collected $32 billion in tax for governments), yet there is evidence that they also collude with smuggling where it suits them—about 10% of world cigarette trade is smuggled but in some countries it reaches 50%.

The total annual world production of tobacco is about 6 million tonnes and the *Food and Agriculture Organisation* (FAO) predicts an increase in annual production and in the number of smokers. The high taxation on tobacco means that it is widely smuggled as evidenced by official figures. In 2003, the number of cigarettes officially recorded as being exported (851 billion) exceeds the number officially recorded as being imported

(664 billion)! The total world cigarette sales in 2003 were $340 billion and the lost tax revenue was estimated to be $40 billion.

Opium—the good, the bad and the ugly in one plant

> *It banishes melancholy, begets confidence, converts fear into boldness, makes the silent eloquent, and dastards brave.*
>
> *—John Brown (influential Edinburgh physician, published Elements of Medicine, 1795)*

Opium is the term given to the alkaloid-rich material derived from the opium poppy, *Papaver somniferum* which grew in parts of south-east Europe and western Asia. The opium poppy is quite tolerant of the growing conditions and the plant is now cultivated worldwide. The major narcotic is morphine, a chemical that has wonderful beneficial properties for some humans and terrible long-term effects on others. Strangely, the very long history of morphine use suggests that some societies can gain the benefits of morphine use without the negative effects of the drug yet other societies seem to find morphine abuse a serious problem.

Thousands of years of use begins

There are several European Neolithic sites where seeds of the opium poppy have been found, with one site in Spain having been dated to 4200 BC. There is evidence that by 3400 BC the Sumerians in Mesopotamia (Iraq) were cultivating the plant, which they called 'Hul Gil' (joy plant). The Assyrians, Babylonians and Egyptians continued to exploit the plant and a trade was developed in opium to ports in the Mediterranean.[9] Opium was used in religious ritual, medicine and what currently would be called recreational use.

The Greek writers make numerous references to opium with some evidence that they used species other than *P. somniferum*. Hippocrates, considered by some to be the father of medicine, rejected the supernatural attributes of opium but acknowledges opium's usefulness as a narcotic, especially in the treatment of certain diseases. It has been claimed that Alexander the Great took opium to Persia and India in 330 BC. The Romans continued the use of opium in medical applications and there is little evidence that addiction was a problem in these ancient cultures. After the rise of Islam, the use of opium in medicine was further developed and documented. At some stage, the Arab spice traders took opium to the Far East and some credited them with introducing it to China. By the middle ages, the drinking of opium mixtures for recreational use is recorded in Persia and India.

The opium trade—hundreds of years of abuse

Given that the opium poppy can be grown in many countries and opium resin for smoking can be produced with very little technical knowledge or expensive processing, one would expect opium to be priced very much as a commodity product, with limited opportunities for the generation of wealth for those who make or trade in the substance. This was indeed the case up until the seventeenth century and it might have remained so, except that a new group of consumers were found and government prohibition in some countries forced the price of opium up. Those countries which had few controls of opium use saw little increase in price to consumers until each country, one by one, made opium use illegal in the late nineteenth century or early twentieth century. This is yet another example of the way in which governments play a large part in creating the conditions for wealth generation associated with the NP use.

Although opium had been introduced to China as a medicine by Arab traders in the eleventh century, its use was very restricted when the first European traders (the Portuguese) began to supply it in 1557. From 1637 onwards, opium became the main commodity of British trade with China and it was in 1700s that the Dutch introduced to the Chinese the famous practice of smoking opium in a tobacco pipe. It seems that such opium use began in the Dutch colonies and then spread to mainland China via Taiwan.[14] The Chinese authorities, realising the harm being done to many of their citizens on the East Coast, tried to control this trade. In 1729, the Chinese Emperor, Yung Cheng, issued a decree prohibiting opium, except for medicinal purposes, but the extraordinary euphoric effects of the narcotic were well known and imports continued. Indeed, by 1753, the authorities were taxing this illegal import. By the end of seventeenth century, Chinese Emperor Kia King banned opium completely, including the cultivation of the plant. However, this abolition of opium production in China opened up an opportunity for others to step in and supply the drug. Large profits were to be made by trading opium with the Chinese and traders from Europe, North America (who bought their opium from the Ottoman Empire) and Japan, all took advantage of this growing business. The British exported 60 tonnes to China in 1776, 300 tonnes in 1790 and 1500 tonnes in 1830. The East India Company, knowing the harmful effects of sustained opium use, banned its ships from carrying opium but they continued selling the opium they were making on an industrial scale in Calcutta, India to others and allowing them to trade with the Chinese. The British government had begun taking a direct interest in the affairs of the East India Company because of the economic importance of the company to the British economy. Indeed, the British Prime Minister, William Pitt, knew of this terrible trade but was so worried about the loss of silver bullion to pay for tea imports from China should the opium trade between India and China decline that he lobbied for it to continue. Despite the banning of the use of opium by the Chinese government, many Chinese gangs and individuals were making money from opium dealing and the illegal trade increased. In 1838, the Chinese Emperor, Tao-kwang, appointed Commissioner Lin Tze-su to stop the opium trade. Lin took the drastic action of setting fire to warehouses and the British hulks in port which contained the opium. He also arrested some British citizens. Outraged, the British shelled Canton in a punitive response. By 1840, China and the United Kingdom were at war but the superior military technology available to the Britain made the struggle an unequal one. The Chinese signed a humiliating peace treaty in 1842, paying the British a large sum of money and giving the British the control of Hong Kong and also access to other new open ports (Shanghai being the most notable, where a sizeable European population was to be found 100 year later before they fled the Japanese). A second opium war (1856–60), in which the French and the United Kingdom defeated China again, allowed even greater access for European countries to many parts of China. Further conflict at intervals throughout the nineteenth century gave the Chinese an understandable, negative view of the advantages of trading with the Europeans and Americans. Opium imports to China represented about 16% of the total imports to that country in the nineteenth century. Even at the start of the twentieth century, addiction in China remained a major problem. In 1906, of the 41,000 tonnes of opium produced, 39,000 tonnes were consumed in China.

It is hard to overemphasise the impact of the consumption of opium on China in the nineteenth century. Before this trade began, China was a proud, self-sufficient, technologically and scientifically advanced nation (some have claimed China was often 4–7 centuries in advance of the European nations in these respects). By the end of the nineteenth century China was weakened to the extent that it was hard to govern, impoverished and technologically surpassed by its neighbour Japan, North America and the European nations. The roots of its troubled history in the twentieth century could be said to lie in the soil of Bengal where the East India Company grew its opium.

The Opium War, also called the Anglo-Chinese War, was the most humiliating defeat China ever suffered. In European history, it is perhaps the most sordid, base, and vicious event in European history, possibly, just possibly, overshadowed by the excesses of the Third Reich in the twentieth century. (Richard Hooker, 1999)

The medical use of opium and morphine

Opium was used medically in all cultures where it was known. In the fifteenth century, the great physician Paracelsus was so impressed by the medical potential of opium that he devised a special mixture, Laudanum (after the Latin word for praise), which included opium. There were many others making mixtures containing opium, one of the best known came from the English physician Thomas Sydenham, who in the 1660s devised a cordial rich in many NPs but with morphine as its major active ingredient (0.5 litre of sherry or other fortified wine, to which were added saffron, cloves, cinnamon and 50 g opium). The use of Laudanum as a sedative helped the depressed, the restless of all ages (it was given commonly to children in the nineteenth century) and it dampened pain. However, there was also a growing recreational use among all classes. In the newly industrialised towns in England, many workers saw opium as a cheaper, more effective way of escaping reality than alcohol. Among the literati, opium was also a popular form of escapism and considered to aid creativity. Many of the great nineteenth century English writers were opium users, including Lord Byron, Samuel Taylor Coleridge, Percy Bysshe Shelley, Elizabeth Barrett Browning, Charles Dickens, Lewis Carroll, Edgar Allan Poe and John Keats.

The medical use of opium was changed dramatically after the introduction of some of the purified ingredients extracted from the poppies. In 1803, the German pharmacist FW Serturner isolated the principal alkaloid in opium, which he named morphium after Morpheus, the Greek god of dreams. A little later, two more alkaloids were isolated from opium, codeine (1832) and papaverine (1848). By the 1850s, the medicinal use of pure alkaloids, rather than crude opium preparations, was common in Europe. Some extolled the safety, reliability and a long-lasting effect of morphine and considered it as 'God's own medicine'. However, it was possibly modern warfare, with mass produced, rapid firing and accurate weapons, that really stimulated the use of morphine. For example, in the United States the drug was widely used during and after the Civil War because it so successfully controlled the pain of wounded soldiers. But this early mass use made the addictive nature of morphine use even clearer and in the United States, morphine addiction became known as 'the army disease' or 'soldier's disease'. The Crimean War in Europe also stimulated use, as did the First World War and all subsequent wars. However, the evident morphine addiction in a group of men which society held as worthy (i.e., soldiers) prompted a scientific search for a potent, but non-addictive, painkiller. The Bayer Pharmaceutical Company of Germany was the first to produce, by the acetylation of morphine, a new drug under the brand name *Heroin* (Figure 2.8). Sadly, although initially thought to be non-addictive, subsequent studies showed heroin to have narcotic and addictive properties far exceeding those of morphine.

Currently, the world spends billions of dollars per year trying to stop the illegal production and use of opium, yet all societies greatly value morphine as one of the most effective painkillers available for those needing pain relief. Annual world production of opium is currently estimated at 5000 tonnes (about 10% of the production 100

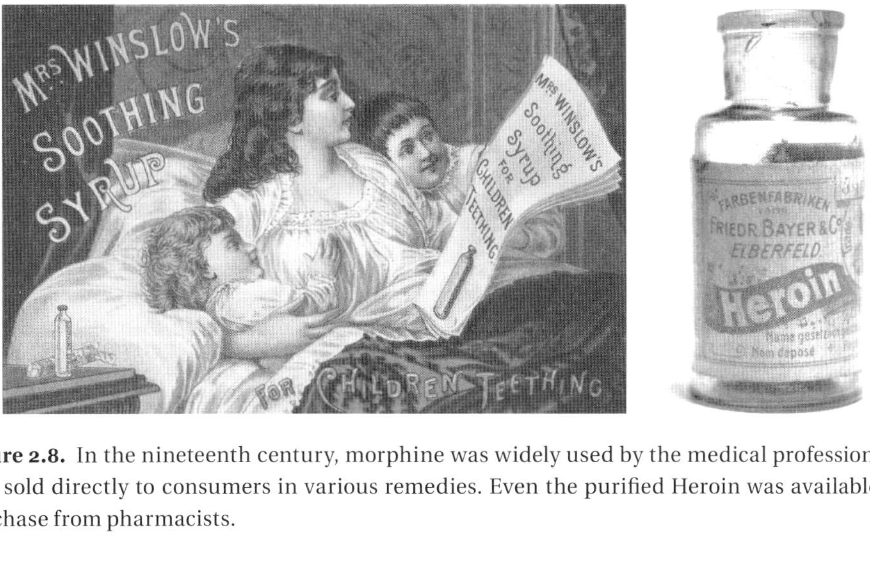

Figure 2.8. In the nineteenth century, morphine was widely used by the medical profession but also sold directly to consumers in various remedies. Even the purified Heroin was available for purchase from pharmacists.

years ago) with only 200 tonnes per annum used for legal medical purposes. Like all attempts to control the use of any NP, prohibition has simply enriched many criminals and made criminals of some people, who have chosen to enrich their lives by limited opium use. The reasons why it is so hard to control the use of any NP is perfectly illustrated by opium. In 2002, an Afghan farmer growing opium would receive $300 per kg, hence make about 10 times more profit per hectare than growing wheat. Local dealers in Afghanistan would expect to receive $800 per kg yet the street price in Europe is equivalent to $16,000 per kg. After the Taliban government fell, Afghanistan's share of the world market increased many fold. Opium poppy production now occupies nearly 10% of the country's total cropland, supplying over 90% of the world's need and yielding over 60% of Afghanistan's GNP. Even if opium production is abolished in Afghanistan, production will simply increase elsewhere to meet the demands of the market—the area of land needed to supply the world is trivial (<100,000 ha). As has been noted many, many times, unless the demand from the consumers decreases, opium production will be sustained. One interesting question is why opium abuse was so uncommon until recently and why only some people succumb to excess opium use?

Cannabis—a valuable plant or terrible narcotic?

Cannabis is an annual herb that is native to central Asia. The species is known for its use to make the fibre hemp and its use as a psychoactive material. *Cannabis sativa* can be selected to contain minimal amounts of the main psychoactive chemical THC (Δ9- tetrahydrocannabinol), to make hemp, or large amount of THC to consume in various forms as 'hash' (hashish). There is evidence of the use of cannabis (the dried leaves and flowers or resins obtained from them) in medicine, ritual, religion and recreation for at least 3000 years. Its use was tolerated in many regions of the world until the twentieth century when narcotic control laws spread throughout the developed world. Despite these laws the UN reports that 4% of the human population have used cannabis in the past 12 months and campaigns to legalise its use began in the 1960s and it continues. Many sufferers of serious medical conditions report that limited cannabis consumption alleviates their symptoms and research is being undertaken to verify the beneficial effects of limited THC intake for certain patients. There is credible evidence that cannabis consumption is less harmful, and causes less dependence, than the legal substances alcohol and tobacco (Figure 2.9) and this evidence gives politicians a great dilemma. Because the cannabis plant tolerates a wide range of growing conditions (including an ability to thrive indoors when grown hydroponically under artificial lights), its 'domestic' production is global but some countries illegally export significant amounts. The annual global sale of cannabis has been estimated by the UN at $140 billion (with the market in the United States worth $35 billion). In several countries, cannabis consumption was higher in 1980 than in 2000 but recent UN surveys suggest a steady rise in cannabis consumption worldwide.

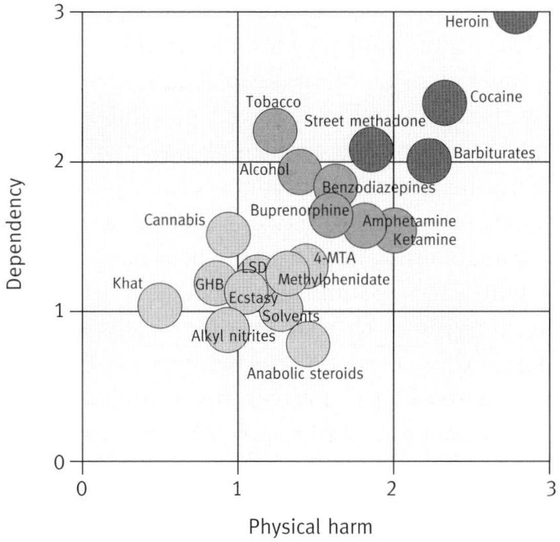

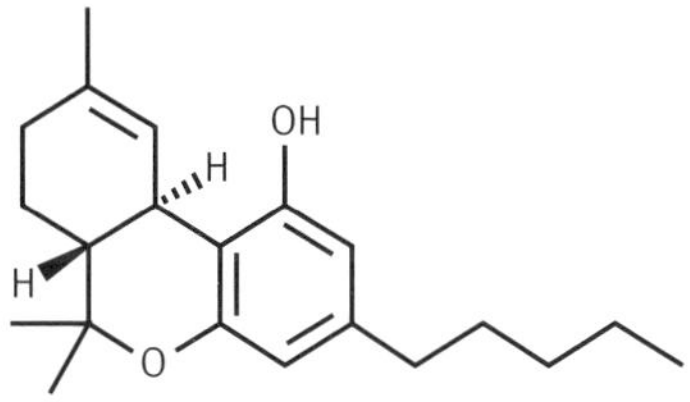

THC (Δ9-tetrahydrocannabinol)

Figure 2.9. An estimate of the detrimental effects of some NP-rich drugs and some synthetic drugs suggests that cannabis (with its important NP THC) is less harmful than some widely used legal substances. (See Nutt D, King LA, Saulsbury W, Blakemore C. (2007). Development of a rational scale to assess the harm of drugs of potential misuse. *The Lancet*, 369, 1047–53.)

Plant and microbial pigments—pretty flowers and colourful mushrooms

Why consider plant pigments as NPs?

Plant pigments are included at this point despite the fact that some are not universally accepted as being NPs. However, as is explained in Chapter 9, the categorisation of substances into neat groups cannot be justified in any meaningful way. In Chapter 9, it is argued that the evolutionary constraints that gave rise to substances which humans have grouped artificially into NPs have also shaped the evolution of plant pigments.

There are hundreds, possibly thousands, of chemicals made by organisms which have molecular structures that result in them absorbing light—hence they are 'coloured' to our eyes. The precise molecular structure of any molecule determines which wavelengths of the visible[18] light are absorbed. Microbes, and subsequently plants, evolved pigments to absorb harmful wavelengths of light, to capture light energy, to act as visual signals to animals after animal behaviour evolved and to use for light sensing. However, it is clear that for some uses, once a particularly effective light absorbing chemical had evolved, the use of that form of pigment was conserved. For example, chlorophyll diversity is quite limited after billions of years. So one has to question why plants as a group make several hundreds of different red/orange/yellow pigments (members of the *carotenoid* family, produced by the *isoprenoid* pathway introduced in Chapter 3, a pathway which makes some important NPs). It seems hard to believe that every carotenoid serves a specific role, so not only do some carotenoids seem to be, like some NPs, optional extras, they are also made by a pathway used to make NPs. Consequently, until we reach Chapter 9 where this relationship is explored further and an explanation given, carotenoids will be treated as NPs.

Similar arguments to those outlined for the carotenoids apply to the phenylpropanoid pigments (see Chapter 3 to remind yourself which NPs are made by this pathway) which account for most of the blue, pink, purple and some yellow and red plant pigments.

The importance of plant pigments to humans

One of the evolutionary selection pressures for the evolution of plant pigments was the evolution of vision in animals. Plants that were coevolving with animals could gain fitness by becoming more obvious to those animals. For example, a plant benefiting from insect or bird pollination gains fitness by having its flowers very visible to those beneficial visitors by being 'colourful'. Likewise, the typical red/orange/yellow pigments in many fruits not only make the fruit more visible (hence attractive) to animals which will disperse the seeds in the fruits but can also signal the 'ripeness' of the fruit (i.e., if the seeds are mature enough) so that the animals learn to take the fruit when it benefits the plant most. Colourful signals can also be used by animals to identify plants, or plant parts, that they best avoid. Interestingly, many colours that we find attractive in animals are actually derived by those animals from their food originally made by plants. For example, goldfish and flamingos need carotenoids in their diets to remain colourful, and egg yolks and butter are coloured by plant pigments.

Given the co-evolution of plants and animals, it is not surprising that humans inherited from their ancestors an acute ability to sense plant pigments—there were rather few non-plant pigments in the human environment until recent centuries.

As in the case of all NP-rich plant products, plant pigments have long been adopted in rituals. The use of flowers in ceremonies is common to many societies—weddings, funerals and many official functions would be dull affairs without some colourful flowers. Likewise, what better way to please someone than to give a bunch or posy of flowers. Once the flowers were locally grown and seasonal but now the well-developed international trade in cut flowers spans the globe. An email or phone call can result in flowers being delivered to a recipient at the other side of the world within hours. The cut flower market has blossomed in recent decades. In the United Kingdom, the fresh cut flower and indoor plant market is worth over £1.5 billion at retail level (to put this in perspective, the UK music industry is worth around £2 billion). Some other European citizens spend 2–4 times the amount the UK citizen spends each year on flowers. However, flowers are not only bought as gifts but also as a way of enhancing one's own home. In the United Kingdom, around 60% of spending represents people buying flowers and plants for themselves.

World sales of flowers were $44 billion in 2002, so it is a large and growing commercial activity globally. Brazil exports over 500 million rose stems annually and other developing countries are entering the market as suppliers. In Ecuador, the industry employs 45,000 people directly and contributed over US $100 million to overall export earnings in 1997. In Colombia, the industry is estimated to employ 80,000 people directly and another 50,000 indirectly. It is Colombia's fourth largest export earner. The biggest wholesale market is in the Netherlands, where there are 100,000 jobs in the flower industry and remarkably 10% of all flowers sold in the United States come via The Netherlands. No doubt the world's flower growers are looking with relish towards China where the average person spends less than a $1 per year on flowers in contrast to the Swiss who lead the world with an annual spend of $112 per person.

Wine and spirits—mainly water with a few wonderful NPs

At the start of the twenty-first century, 250,000 million litres of wine were being consumed annually, valued at about $100 billion. The market is growing by about 5% annually. Wine is a chemically complex mixture but it is not the major constituents (water, alcohol, sugar and organic acids) that give a wine its value but the minor ones—the plant-derived NPs or compounds derived from them. It is the NPs that give a wine a unique flavour or odour; hence wine experts are simply well-trained, capable NP detectors.

Wine making can be traced far back (4000 BC) into recorded human history and might have begun near the Caspian Sea then spread towards the Mediterranean. By 2700 BC, wine was being made in Egypt, where as might be expected of an NP-rich luxury, it was consumed by royalty and priests. The Greeks and Romans improved the husbandry of vines and introduced vines to newly colonised areas. The Greeks improved the ways of making, storing and transporting wine so that a vigorous wine trade was established around the Mediterranean. After the Romans, Christian monasteries were especially influential in maintaining and developing the skills of growing grapes and making wine. As the European explorers spread around the world they took vines and wines with them to new lands. These new vineyards initially satisfied local consumers but in the last half of the twentieth century, the vineyards of South Africa, the United States, Australia, Chile and New Zealand produced high-quality wines which flooded into the European and North American markets. Wine is another example of an NP-rich product where a monopoly was impossible because the vine is so tolerant to growing conditions. However, the growing conditions, combined with the soil properties, both restricted to relatively small areas of the world to produce distinctive wines and a form of branding has developed to increase the value of certain wines. In Europe, the law can give protection to a product made in one geographical area so that only a product made in its traditional location can use a particular regional name for marketing purposes. So a wine labelled as Bordeaux or Burgundy must have been made in the respective region and is effectively branded to achieve a higher value than its NP composition might justify.

Predictably with an NP-rich product like wine, consumption is accompanied by ritual. Wines are served at official dinners, at weddings, to celebrate births and to show displays of triumph after car racing. Special drinking vessels (wine glasses), special storage containers (wine bottles), special storage racks and wine cellars, all point to wine having a special value in European culture.

Some spirits and many liquors are also NP-rich products, some markedly so. Thus, gin gains its special flavour from the NPs found in juniper berries. Some companies developed more distinctive flavours in their brands of gin by adding other NP-rich 'botanicals' (lemon, orange, anise, angelica root, licorice, cinnamon, coriander and cassia).

Beer—the wonderful taste of the NPs found in hops

The fermentation of sugar to produce alcohol was discovered by many societies. The oldest physical evidence for the production of beer is in the remains of 5000-year-old pots from Iran. The source of the sugars for making beer depended on the plants available and in Europe the main sources were starch found in grains grown for food—wheat, barley, oats and rye. By allowing grains to germinate for a few days, the starch in the grain is broken down by enzyme action to produce soluble sugars. The simple techniques needed to produce an acceptable brew made beer a very widely available drink and it had the advantage of being alcoholic and usually free of pathogenic microbes because the brewing process often involves a boiling of the 'mash'. In many parts of the world it is safer to drink beer than tap water.

The NP-rich hop plant (*Humulus lupulus*) was first cultivated in Germany in the eighth century and its use in brewing was first recorded in 1067. In the following centuries, the use of hops to flavour, and help preserve beer, spread throughout Europe (reaching Britain in the early sixteenth century) and then was taken globally by the explorers. Extracts of the hop inflorescence provides a bitter taste that complements the sweetness of the remaining soluble sugars in the brew (which have a malty taste). It is also thought that some of the NPs in hops inhibit microbial growth, hence reduce spoilage which could be a problem in the small local brewhouses before industrial brewing emerged.

More than 250 chemicals have been identified in hop oils and no doubt more minor products remain to be characterised. Annual world beer sales are currently about $300 billion (130 billion litres).

Soft drinks—a very profitable way of selling water by adding a few cheap NPs

Until recently drinking water was a product that commanded a low price in most societies. How can one make lots of money by selling water? Simply by exploiting the human desire to experience the exotic tastes and smells associated with NPs. The first recorded NP-flavoured drink was lemon-flavoured water marketed by Compagnie de Limonadiers, a company given a monopoly to sell that product in Paris. NP-flavoured drinks were often regarded as inherently healthy for two reasons. First, because fruits were regarded as good, nutritious foods (we now know that fruits are a good source of vitamins). Second, the common, simple way of extracting the flavoursome NPs from the plant material was to boil the plant material in water and that killed the pathogens that were common in drinking water supplies in towns at that time. So such drinks, if freshly made, were a good choice for the consumer.

The next major development in soft drinks was the concept of carbonation—making the water fizzy by pressurising the water with carbon dioxide gas. This was first achieved by the radical preacher and scientist Joseph Priestley in 1767. Within a few years, others had developed various types of apparatus to generate the carbon dioxide and to carbonate drinks for sale in shops. The mass market awaited the development of ways of economically making glass bottles, bottle caps and bottling machines which were achieved in the nineteenth century. The introduction of ginger beer (1851), root beer (1876), Dr Pepper (1885), Coca-Cola (1886) and Pepsi Cola (1898) were examples of particular NP-based mixtures finding large markets. The development of metal cans in the twentieth century made the soft drink businesses some of the most successful global brands. The range of NPs used in drinks is now very large but many synthetic versions are used, although the preference of the more wealthy consumers for 'natural' flavours sustains the market for some major NPs. Some soft drinks also appeal to consumers because they contain NPs that have a physiological effect such as caffeine and theobromine. Such chemicals are mildly addictive which helps sales.

The world market for soft drinks is about 500 billion litres. In 2002, the world sales of soft drinks reached $200 billion, with the US market accounting for nearly one-third.

Khat

Catha edulis produces cathenone, an alkaloid with a mild stimulatory effect, thought to be similar to amphetamine. There has been a debate as to whether the plant originated in Ethiopia and was then taken to Yemen or vice versa. The leaves of the plant are chewed. The ancient Egyptians considered the plant a divine food and it has been used for thousands of years in the regions of the Horn of Africa and East Africa. Its use has spread slowly but the need to chew large amounts of the leaf and the habit of users to expectorate frequently means that in western societies it has never gained popularity. However, Somali immigrants to some countries have spread the habit globally and a small international trade has begun. In many countries Khat is regarded as a narcotic but in Britain, surprisingly (and somewhat inconsistently) it is still legal to use it.

Betel nut

The nut of the palm *Areca catechu* contains a mildly stimulatory alkaloid and in some Asian countries the chewing of pieces of the nut is enjoyed because of the mild euphoric state it induces. Restrictions in the use of betel nut remain rather few despite some evidence that its use could be harmful. As with other plant products that are chewed, the expectoration that heavy users experience is an annoyance to non-users and that is likely to limit its use in many societies.

Perfumes and scents—NPs oldest human obsession?

Flowers produce a range of volatile chemicals, some of which are attractive to some organisms, some repellent and some which are unlikely to be detected by any organism (for reasons which are discussed in Chapter 9). Most humans find the scent of some flowers attractive and they enjoy the sensation that detection of these smells gives them. The selective breeding of plants bearing attractive flowers has resulted in some cultivated species which have extremely strong and very characteristic odours. Many people can identify roses, lavender, hyacinth, *Louisiana* iris or lily of the valley by their smell alone and extracts of these flowers have been important to the perfume industry for centuries. Until the twentieth century, extracts containing NPs were the basis of the perfume industry but once chemists had developed methods to isolate, chemically characterise and synthesise the individual characteristic chemical components of the scents, synthetic chemicals were increasingly used in perfume production and also in food flavourings. However, the last quarter of the twentieth century saw many wealthy consumers express a preference for 'natural' NPs (see Chapter 4 for a discussion of this aspect of human behaviour); hence the market for naturally occurring flavouring and odour chemicals remains healthy. The annual fragrances sale worldwide is about $15 billion.

What does this chapter tell us about how science works?

The wealth that can be accumulated by trading in NPs (and it is noteworthy that actually growing the NP-rich plants seems never to have enriched farmers to a similar extent) not only allowed the arts to flourish but also seems to have encouraged scientific endeavour. The periods of great scientific achievement associated with various regions of the world seem correlated to a degree with the periods of NP-associated wealth generation. However, some very influential scientists have suggested a more direct association between NP consumption and scientific creativity. Charles Darwin felt 'most lethargic, stupid and melancholy' after giving up tobacco for a month. Albert Einstein noted 'that pipe smoking contributes to a somewhat calm and objective judgement in all human affairs'.

Given that this chapter shows how important NPs have been throughout history, the chapter has tried to bridge the sad gulf between the study of science and other disciplines. How many students of science appreciate the importance of NPs? How many students of history appreciate the role that a human obsession with NPs has played throughout history? Science is about understanding the natural world and surely the role of NPs in evolution and in human affairs must be part of science.

Conclusion

Ask a scientist to produce a list of the important biological topics that need to be understood to appreciate human affairs, historically and currently, and it is doubtful if many would mention NPs. Yet surely this chapter shows that the human desire to access a few NPs has been extremely important in human affairs. The language some nations speak, the cultural traditions they follow, the religions they practise and even the sports they play can often be traced to that nation's historical links to the NP trade. This massive impact of NPs on human affairs makes it all the more remarkable that very few biologists are aware of the way humans seem in thrall to NPs. The impact of NPs on the lives of every person, every day is seemingly invisible to most members of society, including scientists. This tells us that scientists do not always have powerful abilities to observe, question and analyse.

3

The Main Classes of NPs—Only a Few Pathways Lead to the Majority of NPs

Research is to see what everybody else has seen, and to think what nobody else has thought.

—*Albert Szent-Györgi (1893–1986)*

Summary

One current estimate of NP diversity totals 170,000 different structures, yet this huge chemical diversity is generated from only a few biochemical pathways that branch from the metabolism shared by most organisms. About 60% of the known NP diversity comes from one ancient pathway (the isoprenoids or terpenoids), another 30% comes from some other ancient pathways related to each other (the polyphenols, phenylpropanoids or polyketides) and less than 10% of NPs (alkaloids) comes from a more diverse family of pathways. There seems to be a rough correlation between the number of species possessing one pathway and the total diversity of NPs made by that route. Consequently, the minor groups of NPs that comprise less than 1% of the total NP diversity (e.g., the glucosinolates) tend to be restricted to a small number of species.

A feature of each of the major NP pathways is that, after a branch point leading from the basic cell metabolism, only a few enzymes are needed to elaborate the few basic carbon skeletons characteristic of that group of substances. It is the subsequent additions, deletions and changes of the basic carbon skeletons that generate the great chemical diversity which characterises NPs. Although each major pathway has its own characteristics, each shows evidence of evolving to generate chemical diversity at low cost. For example, in the case of the isoprenoids, with an echo of the way in which huge protein diversity can be generated by joining together a few building blocks (amino acids) in various ways, a few isoprenoid precursors are made which are then joined together in multiple ways to give a few branches to the pathway. In the case of the polyketides and the glucosinolates, an iterative process produces various chain lengths of related structures depending on the number of cycles operated. The fact that there are so few pathways leading to NPs, and that each shows characteristic cost-saving strategies, will be used in Chapters 5 and 9 to construct an evolutionary argument concerning why NP metabolism was shaped by selection in this way.

Understanding molecular structures led to an understanding of biochemical pathways

In science, the growing understanding of a natural process is characterised by the assembling of data into patterns or groupings, so that a narrative becomes possible. Some scientists concentrate on gathering data, others use their talents to assemble the data into patterns and some do both. However, in biology, a few ask how and why such patterns have arisen because for the past 150 years the concept of natural selection has provided a rule book to help interpret biological processes. One can see this historical progression in the study of NPs. Initially, chemists were content to simply gather data about the structure of individual NPs, then chemists and biochemists began to place the individual NPs into groups, on the basis of the carbon skeleton of the structures and finally questions were asked about why NP metabolism was shaped as it is.

Very soon, after it became possible to determine the elemental composition of naturally occurring chemicals in the eighteenth century, it was noted that such substances commonly contained carbon (C), hydrogen (H) and oxygen (O), and a smaller number of substances also contained one or more other elements—nitrogen (N), sulphur (S) or phosphorous (P). Simple classifications based on elemental composition became possible—one could group chemicals by carbon number, whether they contained N or whether they contained S. But such a classification provided few insights. However, by the middle of the nineteenth century, chemists began to have ideas about the way in which the atoms might link together. It had been observed that the relative proportion of each element in a molecule seemed to be governed by some rules. For example, certain combinations of carbon, oxygen and hydrogen were never found. This led to the development of the concept of *valency*—each element has a characteristic capacity to link to other atoms. Some types of atoms (e.g., carbon) seemed to have a capacity to simultaneously associate with several other individual atoms, but other atoms such as hydrogen simultaneously link only to one other atom. The valency rules explain why only certain combinations of elements were found in substances. The challenge was then to provide an underlying explanation for the valency rules. Many chemists contributed to these developments, but the German August Kekule (1829–96) and the Scot Archibald Scott Couper (1831–92) independently proposed how to represent valency on paper and provided an elegant tool that enabled chemists to record, and share, their thoughts about the structure of molecules. Typically, chemists sought to build up a picture of the way in which the carbon atoms were joined (*the carbon skeleton*) and then they postulated how the other elements were joined to that skeleton. It soon became apparent that certain properties of chemicals made by organisms were a result of characteristic additions (substituent groups) to some common carbon skeletons. Furthermore, knowledge about the properties of a chemical enabled one to predict parts of the structure. For example, the group of naturally occurring acidic chemicals discovered by Scheele, such as lactic acid, tartaric acid, malic acid and few others (see Figure 1.2) shared the –COOH structure, but each type of molecule had a different carbon skeleton to which that group is attached. The fact that shared characteristics were indicative of at least one shared structural feature gave synthetic organic chemists a simple classification scheme to bring an order to their subject—alcohols, organic acids, esters, aldehydes, ketones, phenols, hydrocarbons and so forth. These broad classifications are still used. However, this was only one way of classifying chemical structures. An alternative classification system was based on the underlying carbon skeleton. Linear sequences of linked carbon atoms (aliphatic) shared some properties that distinguished them from sequences of carbon atoms linked in one or more rings (cyclic and aromatic). By the start of the twentieth century, these classifications had brought order to synthetic organic chemistry, and soon attempts were made to tackle the much harder job of classifying naturally made chemicals.

Placing biochemical pathways on metabolic maps

Chemists trying to understand the chemical structures being made by organisms, the physiological chemists as they were then called (see Chapter 1), recognised at an early stage that their job was to 'map' the ways in which organisms could convert one structure into another.[1] Soon the concept *of metabolic pathways* (the trunk roads on the map) gained wide acceptance and many naturally made chemicals were classified on the basis of their location on one particular pathway rather than (or as well as) on their individual chemical structures. As the individual enzymes contributing to the main biochemical pathways were discovered, usually in microbes because they were more readily manipulated experimentally, two-dimensional metabolic maps were built up, where each enzyme was assigned a role in carrying out one conversion (Figure 3.1). Predictably, most biochemists seeking to contribute to the building of the growing metabolic map were working on the biochemical pathways that are common to a majority of organisms, pathways that are needed to build and maintain any cell.[2] These biochemical pathways must have evolved billions of years ago, with each enzyme embedded in a well coordinated, highly evolved network and mutually dependent on other enzymes (see Chapter 9). One characteristic feature of this form of metabolism is the existence of metabolic cycles where individual enzymes can contribute to a series of steps that make up an endless cycle, with branch points into and out of the cycle at some stages. By the middle of the twentieth century, the metabolic map for the basic metabolism common to wide groups of organisms was well established, and subsequent research largely refined it. However, when biochemists began to probe the way in which individual NPs were made, they faced some significant problems. The carbon flow into NP pathways was usually very low, making such studies harder. The postulated intermediates were harder to synthesise, and the cells rich in NPs were usually scarcer and more difficult to work with. In compensation, it was soon realised that, although the NP composition of individual species varied very greatly, this

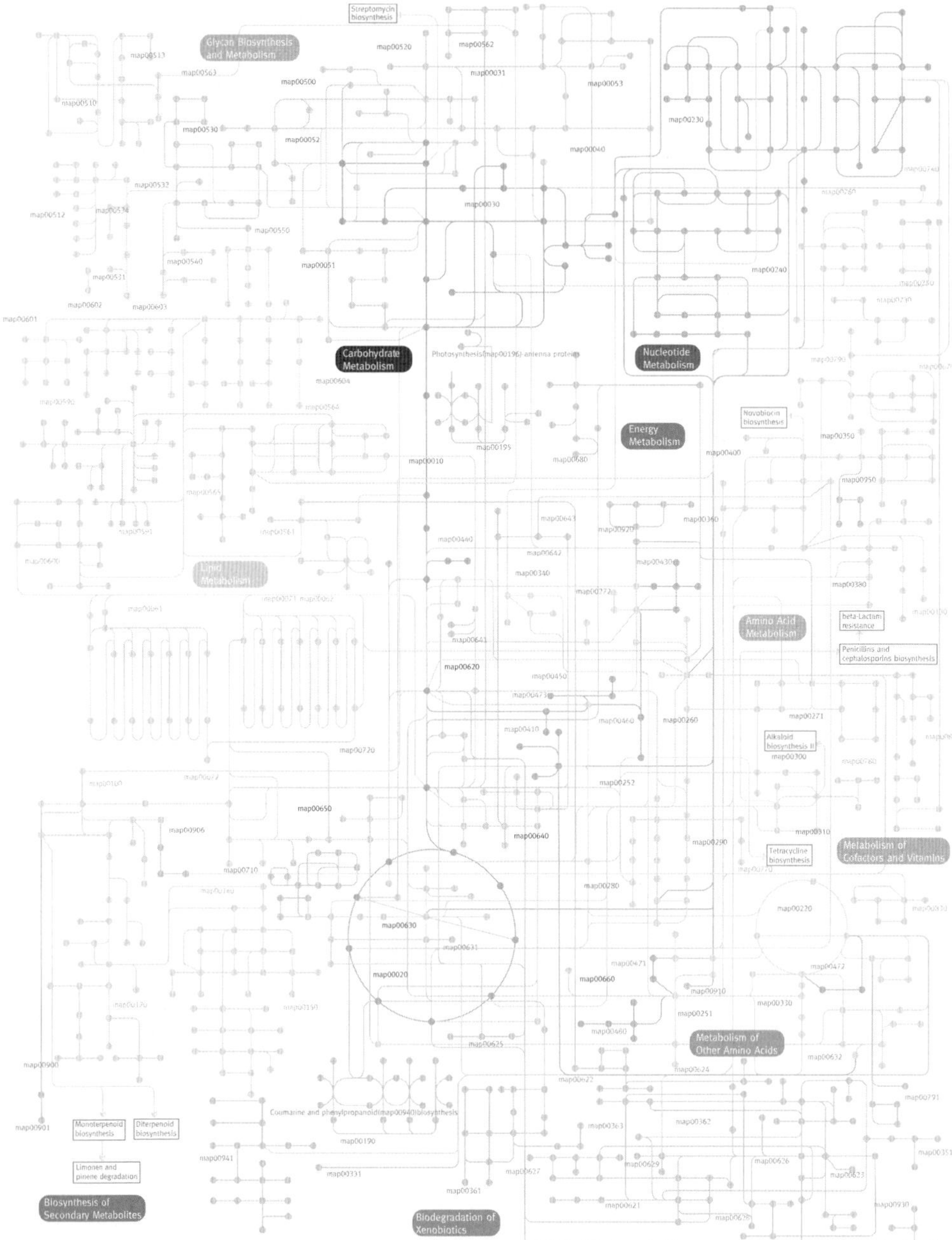

Figure 3.1. Anyone learning biochemistry is introduced to 'metabolic maps' of the type shown here. These 2-D representations of major pathways are becoming increasingly misleading because as this chapter shows the majority of enzymatic transformations in the natural world do not stick to the simple rules implied by such a map. The great majority of chemical diversity made by organisms is made by enzymes that are much more multifunctional so they are less easily placed on a simple 2-D map.

chemical diversity grew from some initial stages of the pathway that were expected to be common to many species sharing that pathway. Thus, researchers working on NP biosynthesis had a common goal when they were studying the first few steps in their chosen major NP pathway even if the later stages of that NP pathway became increasingly species specific. How the basic carbon skeleton of each major NP group was made became the focus of research. Because each major NP pathway must begin with the conversion of some common intermediate (i.e., what Kössel would have called a primary metabolite) to the first dedicated intermediate of that NP pathway, a good strategy was to feed radiolabelled primary metabolites to organisms making NPs and determining whether and how the radiolabel was incorporated into the major NPs. Using this approach, the details of the NP metabolic map began to be drawn.[3]

The major NP pathways

As is explained in Chapter 9, the overall classification[4] of naturally made chemicals into two classes, now commonly known as *primary* and *secondary*, first suggested by Sachs 150 years ago, and subsequently defined more clearly by Kössel 50 years later, has been very unhelpful. There are not two classes of chemicals made by plants and microbes. In Chapter 9, it is argued that some 'NP pathways' contribute to the production of another class of substance, substances that have been selected on the basis of their physicochemical properties.[5] Whereas NPs are selected on the basis of their 'biological activity' (or more precisely their 'biomolecular' properties, as outlined in Chapter 5), some substances made by what have traditionally been considered to be 'NP pathways' serve as colours or as membrane components or play various other roles determined by their physicochemical properties. Hence, the 'NP pathways' are pathways that make NPs, but these pathways are not used exclusively for this purpose. This lack of exclusivity has some interesting evolutionary implications which are discussed in Chapter 9.

The fact that there are tens of thousands of NPs known might seem extremely off-putting when one first approaches NP biosynthesis. All the details and all the 'facts' about the biosynthesis of individual NPs could easily overwhelm those coming to the subject for the first time. Consequently, this chapter avoids as many details as possible (details can be found easily with an internet search) and will try to provide a simple conceptual framework that can be used to see patterns in the details that are provided by others.

What are the key features that characterise NP pathways?

1. There are very few major pathways used to generate NP diversity. One can trace the evolution of two major pathways, which account for nearly 90% of NP diversity, back to their microbial origins.
2. The ancient pathways that make this chemical diversity benefit the producer in at least two different ways; hence, the evolution of these pathways are shaped by multiple selective forces. Some of these chemicals possess potent biological activity while others bring beneficial physicochemical properties to the producer. Consequently, these are multifunctional pathways.

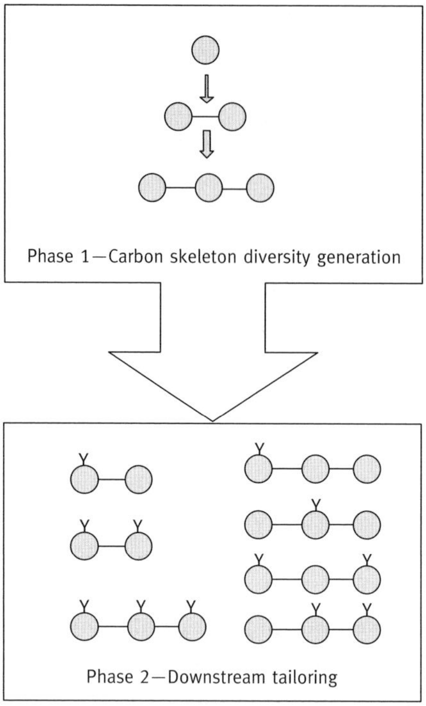

Figure 3.2. The chemical diversity that is characteristic of NPs can be considered to arise in two phases. In the first phase, a few precursors are joined together in a few similar ways (using a modular or iterative processes) to produce families of structures that provide the basic carbon skeletons that characterise the group. In the second phase, enzymes with broad substrate tolerances 'tailor' the skeletons in versatile ways to generate even greater diversity.

3. The major NP pathways can be conceptually divided into two phases. The first phase involves a flexible way of building the basic carbon skeleton characteristic of that family of NPs. The second phase involves versatile ways of making additions, rearrangements and changes to the basic carbon skeleton. Each of these two phases helps generate chemical diversity at low cost, and combining the two flexible phases in sequence amplifies this generation of diversity (Figure 3.2).

The terpenoid or the isoprenoid NPs

Turpentine, the pleasantly smelling piney solvent, gets its name from the Greek word for the tree from which turpentine was distilled. As a solvent, turpentine has medical uses to treat wounds, to combat lice infestation and as an inhalant (the well-known Vick's vapour rub used to contain turpentine). Turpentine is a mixture of 10 carbon hydrocarbons (C_{10}), and the composition varies depending on the tree species (usually a pine or balsam) from which it has been distilled. It was recognised that a family of C_{10} terpenes

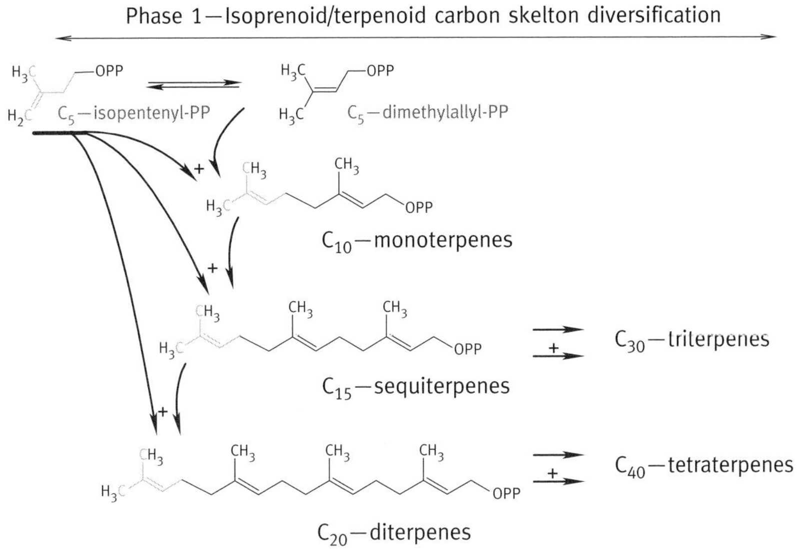

Figure 3.3. The isoprenoid/terpenoid pathways start by joining together C_5 units in multiple ways to generate carbon skeletons made up of multiple C_5 units (e.g., C_{10}, C_{15}, C_{20}, C_{30} and C_{40}). Each of these carbon skeletons can then be tailored to generate further chemical diversity.

were made by many plants; they were part of a larger family of NPs which included chemicals that had carbon skeletons with multiples of C_{10} (e.g., C_{20}, C_{30} and C_{40}). This family of NPs was given the name of the terpenoids, but it was soon realised that the production of the basic C_{10} skeleton involved an earlier joining together of two C_5 carbon skeletons. The backbone of the more fundamental C_5 unit was akin to the hydrocarbon isoprene[6] (C_5H_8); consequently, this pathway also became known as the isoprenoid pathway, and it was proposed that all members of the terpenoid/isoprenoid family were simply made up by repeatedly combining isoprenes. Consequently, the generic formula for a member of this family is $(C_5H_8)_n$, for example, a C_{30} substance had $n = 6$. Although the smaller multiple isoprene backbones were made by adding isoprene units together, larger multiples could be made by adding, for example, a C_{10} and a C_5 to give a C_{15} or two C_{15} units together to give a C_{30} (Figure 3.3). This use of a few basic building blocks to make many different structures is an example of the first phase of an NP pathway where multiple carbon skeletons are generated. This phase is analogous to a modular building system.

Two different ways of producing the precursors feeding into the modular phase of isoprenoids/terpenoids biosynthesis have evolved. One route, the mevalonic acid (MVA) route, makes the basic C_5 building block from the C_5 chemical MVA (confusingly, the substance isoprene is not actually used as the basic building block at all). This route to the isoprenoids takes place in the cytoplasm and is found in a wide range of organisms, including animals. A branch point leading from the basic metabolic pathways takes the three carbon substance acetate to make acetyl-CoA, which is then converted by a series of steps to MVA (Figure 3.4).

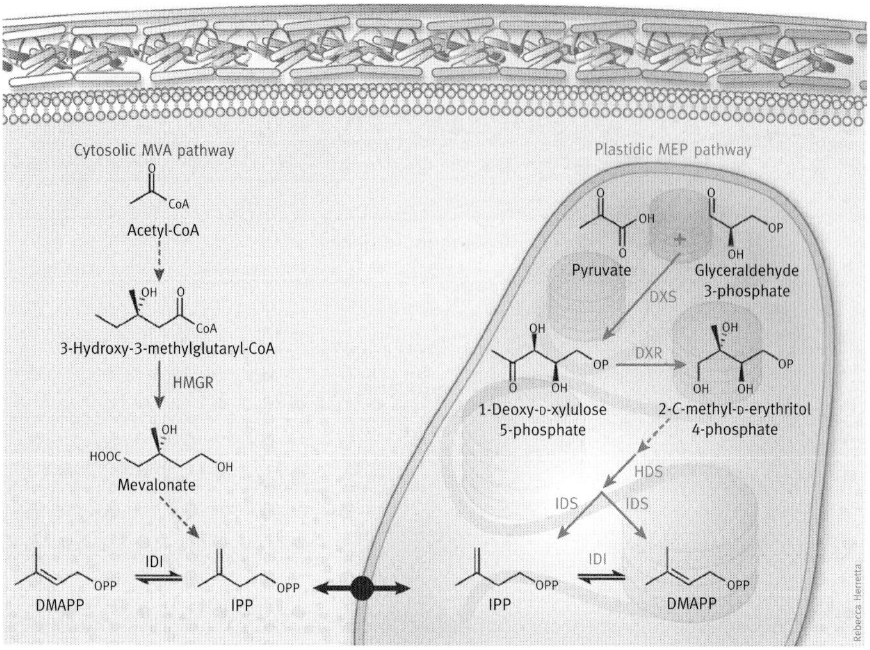

Figure 3.4. There are two routes known for the production of the basic C_5 units used to form the isoprenoid/terpenoid carbon skeletons—the MVA and MEP isoprenoid pathways.

The alternative route, discovered decades later, is known as the MEP (mevalonic acid independent pathway) or the DXP (deoxyxylulose 5-phosphate) pathway. In this pathway, which in plants is located in the plastid and not in the cytoplasm, the C_5 unit isopentenyl pyrophosphate (IPP), which is made in three enzymic steps from MVA in the MVA pathway, comes via five enzymic steps from DXP. In other words, the two isoprenoid pathways start with different initial basic intermediates but converge on the same C_5 unit. The pathways that lead to this point of convergence can be traced back to the microbial past. The MVA pathway possibly evolved in the archaebacteria and the MEP/DXP pathway in eubacteria, with subsequent eukaryotes having inherited their genes for the MEP/DXP route from prokaryotes[7] (Table 3.1). Fungi and animals have only one form of the pathway (MVA), and it seems that while fungi utilize this pathway to generate some NPs, animals mainly use this pathway to provide the substances with beneficial physicochemical properties (which, as is explained completely in Chapter 9, provides an explanation as to why animals, that are not commonly considered as making NPs, retain some of this pathway).

Having produced many different carbon skeletons via the first phase of the isoprenoid/terpenoid pathway, the second phase of diversity generation occurs in all organisms when a number of versatile enzymes carry out a series of chemical additions, rearrangements and deletions by acting on different carbons in the basic skeleton. The

Table 3.1. Evidence that one major NP pathway is ancient and universal but with two different starting points.[8]

Organism	Terpenoid pathways
Eubacteria	MVA or MEP
Archaea	MVA
Seaweed *Laurencia*	MEP
Plants	MVA and MEP
Animals	MVA
Fungi	MVA

URL: http://www.nationmaster.com/encyclopedia/Terpenoid
MVA pathway: In the 1950s, many organisms were found to manufacture terpenoids through the HMG-CoA reductase pathway, which includes mevalonic acid (MVA) as an early precursor. The reactions take place in the cytosol.
MEP/DXP pathway: 2-*C*-methyl-D-erythritol 4-phosphate/1-deoxy-D-xylulose 5-phosphate pathway, first reported in the late 1980s, is also known as non-mevalonate pathway or mevalonic acid independent pathway (MEP). This pathway is located in the plastids of plants and in the structure known as the apicomplexan in protozoa as well as in many bacteria.

enzymes involved in these changes are too many and varied to discuss at this point; suffice to say that the families of such enzymes evolved in microbes and most of these enzymes are characterised by their biosynthetic versatility—a very important clue which is discussed in Chapters 5 and 9 in relation to the evolution of NP diversity.

The polyketide, phenylpropanoid and polyphenol NPs

Microbial fatty acid biosynthesis and the evolution of the polyketides

The second major pathway leading to NPs is another ancient one, and the recent gene analysis of several species reveals that one can trace the evolution of the main parts of this pathway back to their microbial roots. Once again it is useful to look at this pathway as one that has served to provide both substances with useful physicochemical value or beneficial biological activity (NPs). For example, there is now growing evidence[9,10] that the polyketide pathway evolved from the pathway used in bacteria to make fatty acids (substances selected for their physicochemical properties to optimise membrane functioning—see Chapter 9). Fatty acids have long hydrocarbon chains, with each fatty acid having a characteristic chain length, varying from about 4 to 28 carbons in a row but most commonly with 16–22 carbons (Figure 3.5).

The strategy evolved to generate the diversity of fatty acids and not the detailed biochemical mechanisms involved in fatty acid biosynthesis (why organisms make such a diversity of fatty acids is discussed in Chapter 9) is the important point to grasp. It is

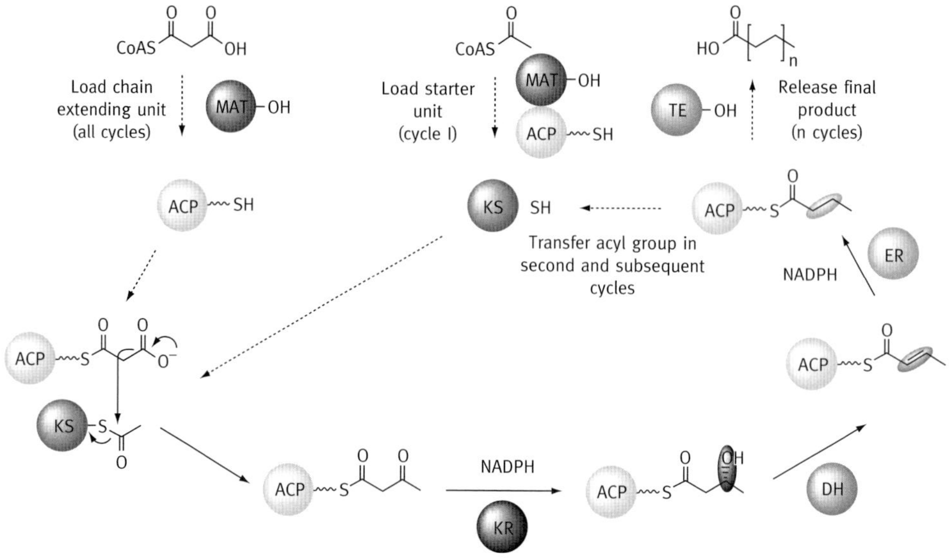

Figure 3.5. Fatty acids are synthesised by an iterative process that adds C_2 units to an elongating chain. The chain length depends on the number of cycles used.

another example of the 'modular' phase of carbon skeleton diversity generation. The hydrocarbon chains of fatty acids are synthesised in a repetitive, or iterative, process with the same basic enzymic steps being repeated each time an extra two-carbon unit is added sequentially to the growing end of a hydrocarbon chain. The number of repetitions determines the final chain length. Consequently, as in the case of the isoprenoid pathway, many related carbon skeletons can be built up using very few enzymes. As in the case of the isoprenoids, the final carbon skeletons can be further elaborated to produce a very wide chemical diversity. What has fatty acid biosynthesis (always discussed in general biochemistry textbooks) got to do with NP biosynthesis (rarely discussed in general biochemistry textbooks)? Well it is now clear that the metabolic properties of cells used to generate fatty acid diversity were built upon in evolution to generate a massive chemical diversity in a group of chemicals known as the polyketides.

The polyketides are a large family of biosynthetically related NPs, some of which have very great pharmaceutical value (polyketide sales total about $10 billion annually, see also Chapter 7). Some antibiotics (erythromycin, monensin, rifamycin), immunosuppressants (rapamycin), antifungal substances (amphotericin), antiparasitic (avermectin) and anticancer drugs (doxorubicin) are polyketides. The term polyketide refers to the fact that the basic carbon skeleton is not a simple hydrocarbon chain as in the case of fatty acids but is a series of linked keto groups in sequence (Figure 3.6). The first phase of this pathway, the generation of carbon skeleton diversification,

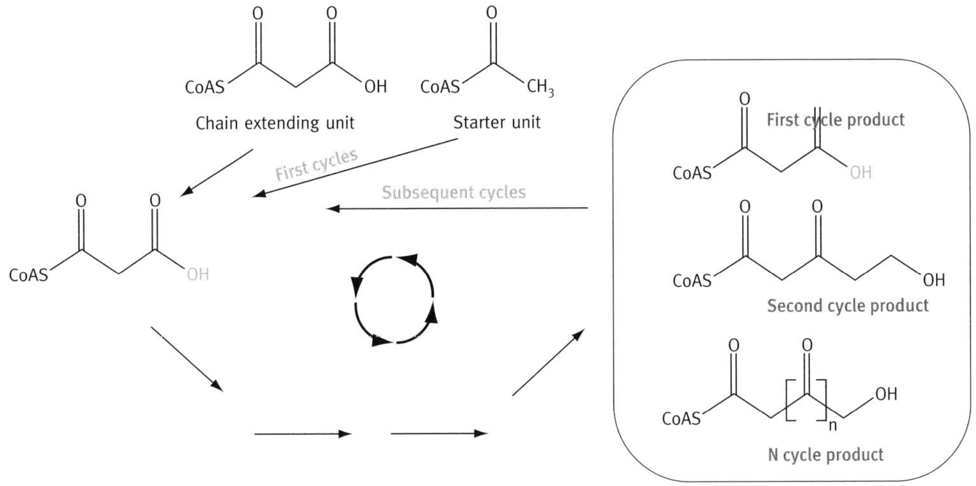

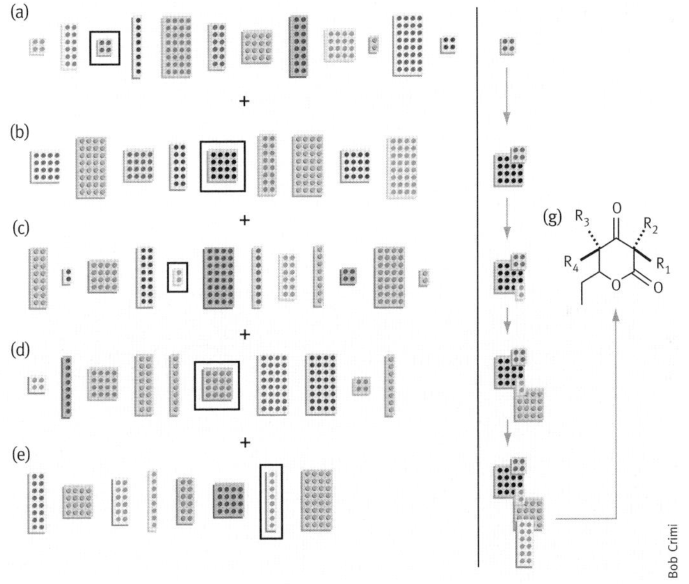

Figure 3.6. Polyketide chains use an iterative process akin to the fatty acid chain elongation process. However, an extra layer of chemical diversity can be generated by feeding different starter units into the system; consequently, the polyketide biosynthetic system has been compared to a 'Lego' system where a few modules can be joined together in many different ways.

generates polyketide chains of various lengths and also incorporates enzymic stages that generate various complex ring structures from these chains. This carbon skeleton diversification phase involves a group of enzymes that function together as a unit or module—the polyketide synthase (PKS) modules. One can identify within this module, a unit that loads the starting material into the module, a unit that elongates the basic carbon chain and a unit that terminates the biosynthetic sequence—the concept is not that different from fatty acid synthesis. By changing the units within a module one can generate molecular diversity at low cost, and it has been proposed that during microbial evolution gene transfer between species allowed units within PKS to move between species.[9,10] By adding or substituting a novel unit into an existing PKS module a rapid, drastic change in carbon skeleton diversification could occur. Furthermore, relatively minor changes in the sequence of the genes encoding for the PKS units can bring about large changes in the carbon skeletons being generated. So not only do the PKS units possess an inherent flexibility but they are always poised to bring about dramatic shifts in the spectrum of polyketides being made. This flexible modular system of building many related structures from a few basic building blocks has been compared to the use of Lego building blocks (Figure 3.6). This flexibility has already been exploited by those seeking to make new polyketides to screen for valuable pharmaceutical properties, with thousands of novel polyketides being created in the laboratory by the simple genetic manipulation of the genes encoding for the PKS units.[11]

The second phase of polyketide diversification by the tailoring is brought about by modifications and additions to the family of polyketide carbon skeletons.

The plant polyphenols—the phenylpropanoids made via the shikimic acid pathway

It was not until a comparison of genomes became possible that the relationship between a major NP synthesis in plants and the polyketide pathway in microbes became evident.[12] In plants, a very large group of NPs were identified as being derived from the shikimic acid. Shikimic acid was first isolated and characterised by Japanese chemists from a plant (*Illicium anisatum*) that the Japanese called shikimi, hence the name (the 'proper' chemical name is (3*R*,4*S*,5*R*)-3,4,5-Trihydroxy-1-cyclohexenecarboxylic acid). The pathway starts with the reaction of phosphoenolpyruvate (PEP) with D-erythrose to give a seven carbon structure (shikimic acid), which is then converted into the aromatic amino acids, phenylalanine and tyrosine. Phenylalanine, a chemical with the C_6–C_3 backbone and which is also used in protein synthesis, can be converted to a range of NPs. The first key enzyme is phenylammonia lyase (PAL), which produces cinnamic acid (another C_6–C_3) from phenylalanine. The generic term for the family of NPs made via PAL is the phenylpropanoids (C_6 = phenyl C_3 = propane). Subsequent enzymic hydroxylations and methylations produce (the second phase of chemical diversity generation) a range of substances—coumaric acid, caffeic acid, ferulic acid, 5-hydroxyferulic acid and sinapic acid. Esterification of the organic acids produces volatile chemicals that contribute to the plant's fragrance (Figure 3.7). Some of the C_6–C_3 substances are linked together to give polyphenolic substances, the best known plant cell wall component lignin (Figure 3.7).

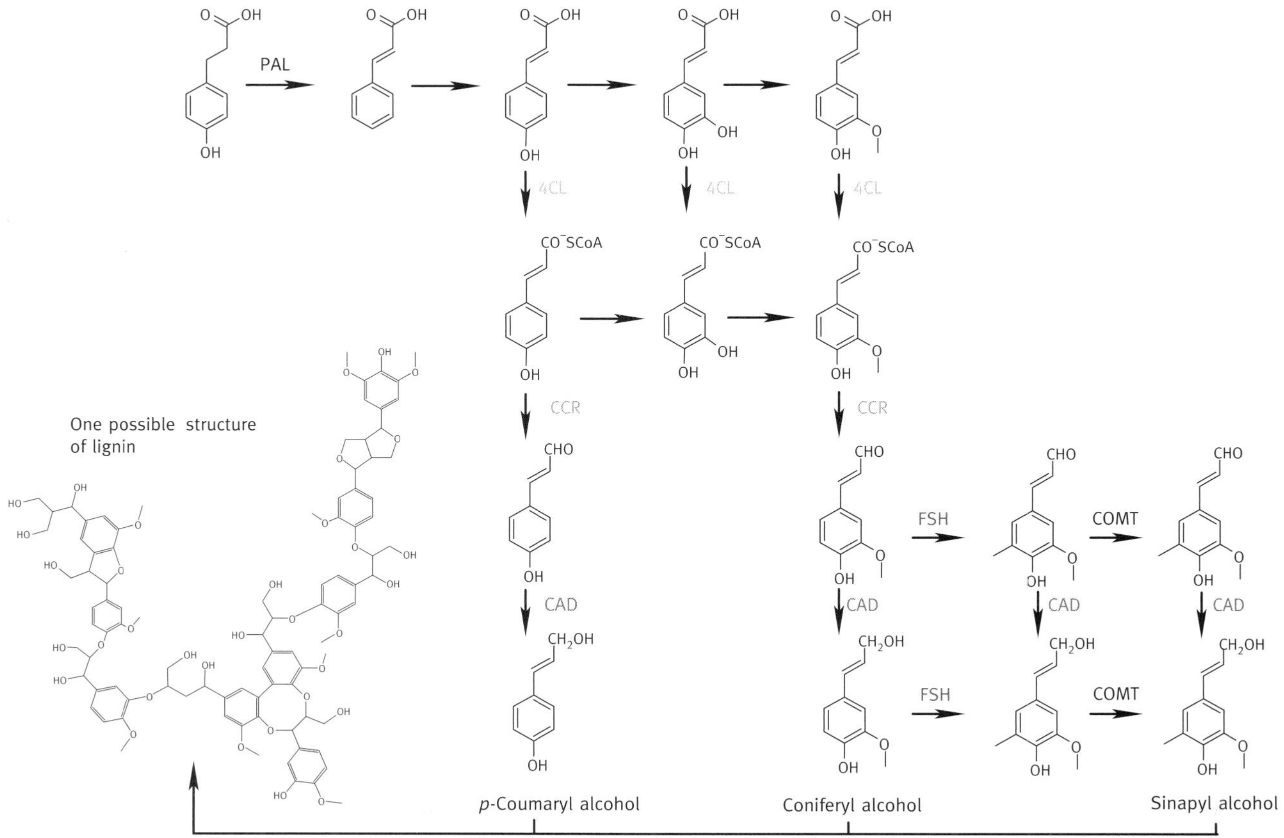

Figure 3.7. The shikimic acid pathway, leading to polyphenols and phenylpropanoids. The basic of C_6–C_3 carbon skeleton is followed by a phase 2 where substituent groups are added to the aromatic ring, or modified, by just a few enzymes. One important role of this pathway is to produce precursors to make the cell wall constituent lignin. Lignin is the second most abundant polymer in the natural world (exceeded only by the other major cell wall constituent cellulose). The structure of lignin is variable; hence that shown here is just an example of the type of structure that might be found.

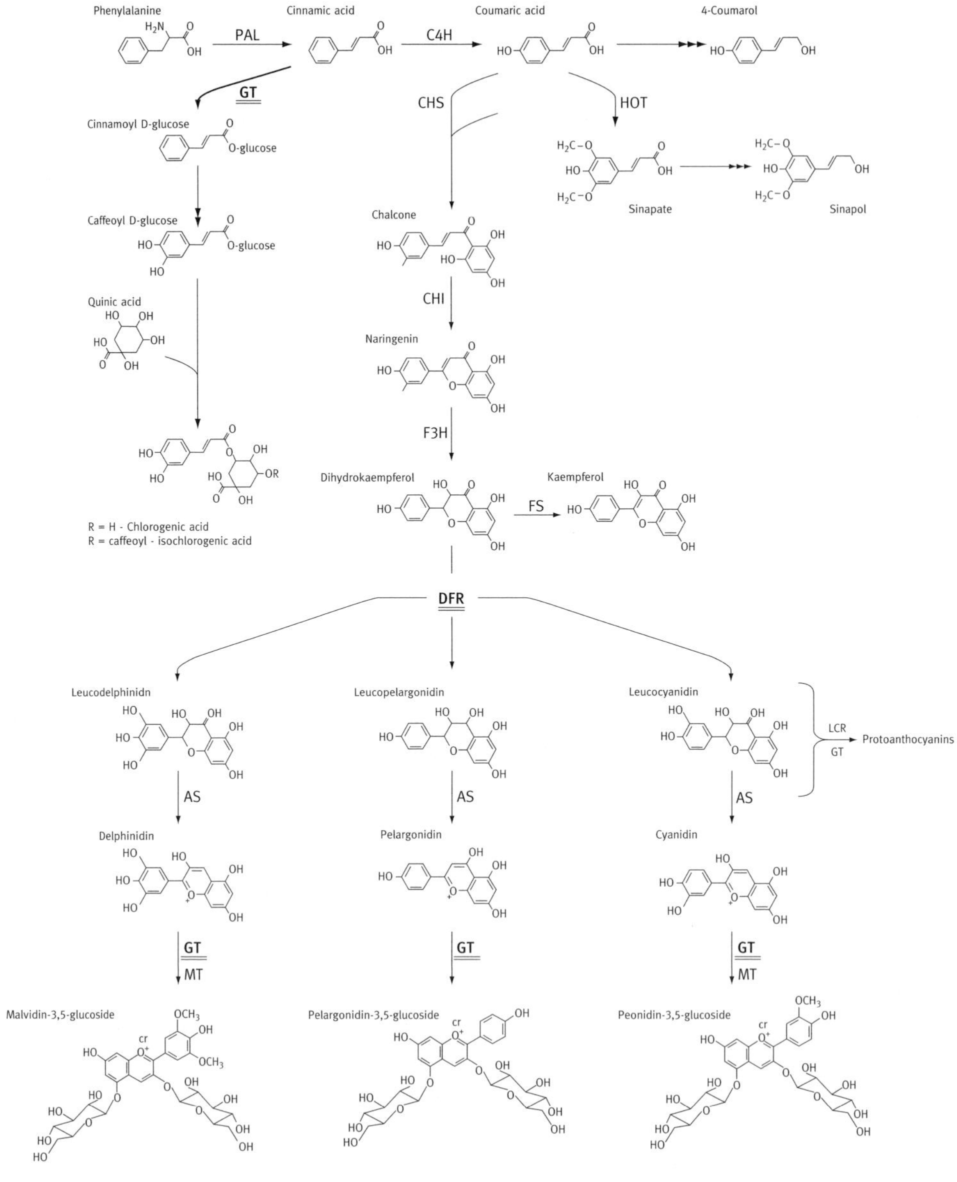

Figure 3.8. The phenylpropanoid pathway, leading to C_6–C_3–C_6 phenylpropanoids. Note the versatile use of some individual enzymes in the phase 2 tailoring. From Aksamit-Stachurska A, Korobczak-Sosna A, Kulma A, Szopa J. (2008). Glycosyltransferase efficiently controls phenylpropanoid pathway. *BMC Biotechnology*, 8, 25–41.

The C_6–C_3 carbon backbone can be extended further by the addition of C_2 units via three iterative acetyl units.[13,14] These units condense with each other and undergo cyclisation to form a new second polyphenolic ring with a final carbon skeleton of C_6–C_3–C_6

(Figure 3.8). The iterative process is carried out by the enzyme chalcone synthase (CHS). The CHS genes are similar to the fungal CHS genes which belong to the type III PKS superfamily (polyketide synthase). So this branch of the phenylpropanoid pathway is analogous to the microbial polyketide pathway (which was itself related to microbial fatty acid synthesis). The new C_6–C_3–C_6 structures produced in this extended first phase of chemical diversity generation is then available to feed into the second phase of structural diversity generation by enzymes that target especially the hydroxyl groups attached to the rings. For example, *O*-glucosyltransferases alone, which can add a sugar molecule to the hydroxyl, can generate over 300 different glycosides of the C_6–C_3–C_6 quercetin.

The alkaloids

The name alkaloid comes from the fact that a number of NPs were found in the nineteenth century which were alkaline. These chemicals were shown to contain a nitrogen molecule, and when their structures were determined it was found that the nitrogen was usually in a heterocyclic ring (a cyclic carbon skeleton with one or more nitrogen atoms in the ring—see Figure 3.9). The N is usually protonated at physiological pH, thus many of these molecules are polar and hence water soluble.

Although far less numerous than the terpenoid/isoprenoid or polyketide NPs, the alkaloids (with an estimated 20,000 different structures)[15] have a special place in NP research because a few are of great value to humans—for example, morphine, theobromine, caffeine, vincristine, quinine, codeine, cocaine, nicotine and strychnine.[16] These often complex chemicals are found in about 20% of vascular plants and a smaller number of fungi, marine invertebrates and a few bacteria.[17]

Given the very great commercial importance of a few of the alkaloids, each of the major alkaloid pathways have been studied extensively producing great detail; a few general principles that concern this chapter have become clear. Unlike the major NP groups discussed so far, the diversity of carbon skeletons found in the alkaloids come not from a versatile iterative or modular synthetic process but by the use of a similar enzymic conversion joining together two different commonly available starting materials. Three specific reactions are found many times in alkaloid biosynthetic routes—the oxidative coupling of two phenols, a reaction of a primary amine (usually derived from a common amino acid) and an aldehyde or ketone and a condensation-type reaction involving three reactants (Figure 3.10). The carbon skeletons of the alkaloids can then be tailored by decarboxylation, aldol condensation, reductive amination or methylation. McKey[18] noted the biosynthetic flexibility of using a few amino acids to feed different carbon skeletons into the alkaloid pathways; he suggested that it would have been relatively simple for evolution to give different plant families related alkaloid pathways by slightly altering the substrate preference of the enzyme carrying out the initial condensation of the amine and keto groups. Although most of the major groups of alkaloids derived from amino acids conform to the simple principles of generation of chemical

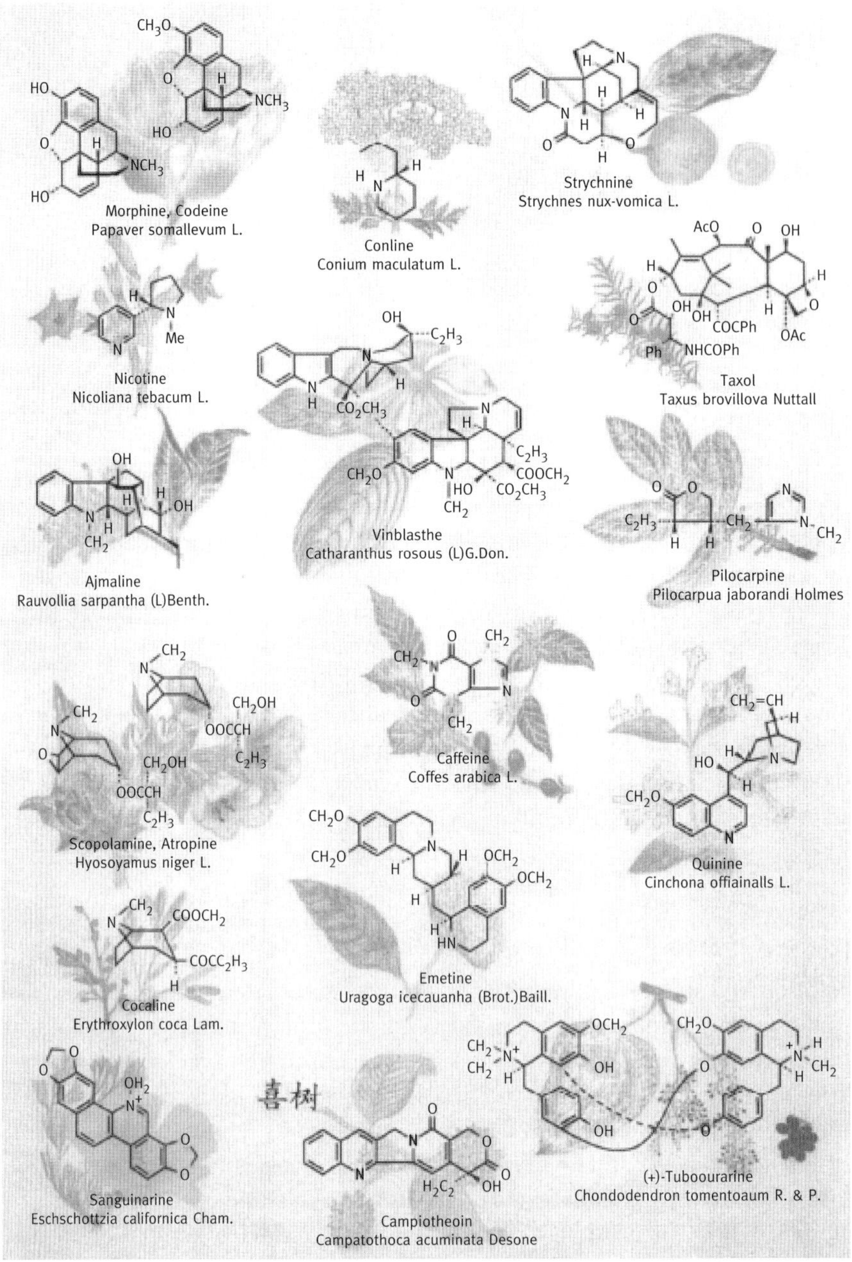

Figure 3.9. Some of the many better known alkaloid structures. From Kutchan TM. (1995). Alkaloid biosynthesis—the basis for metabolic engineering of medicinal plants. *Plant Cell*, 7, 1059–70.

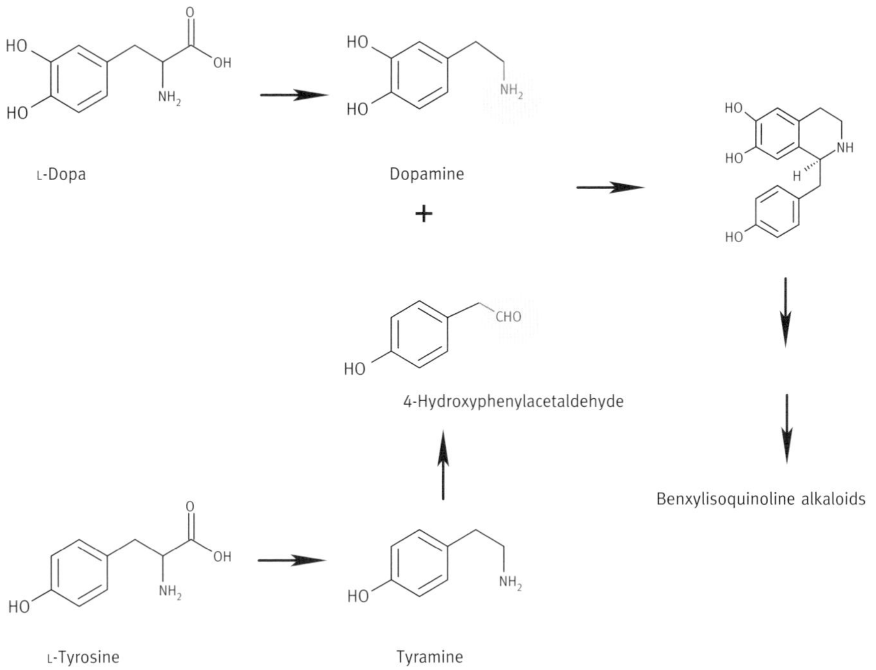

Figure 3.10. An example of a common feature of some types of alkaloid biosynthesis, the linking of a substance with an amino group (such as the amino acid shown here) and a substance with an aldehyde group (in this case, comprising the first stages of the route to the benzylisoquinoline alkaloids).

diversity, there are many individual alkaloids that have evolved to incorporate other NP carbon skeletons as part of their biosynthesis, adding yet more opportunities to generate novel structures.

The glucosinolates

The final class of NPs to be considered are the glucosinolates (ß-thioglucoside-*N*-hydroxysulfates), a class of about 120 chemicals distributed in only 16 plant families. The best known glucosinolates are those that give the characteristic pungent smells and tastes, loved by some but hated by others, of some brassica vegetables such as cabbage, mustard, cress, cauliflower, turnip, brussel sprouts, radish and horseradish. Considerable attention was paid to this group of NPs when it was proposed that some members had anticancer properties. The glucosinolates contain sulphur and nitrogen and are made by the combination of two common substances—glucose and an amino acid (Figure 3.11). The aliphatic glucosinolates are made most commonly from methionine but alanine, leucine and valine can also form part of the basic carbon skeleton.

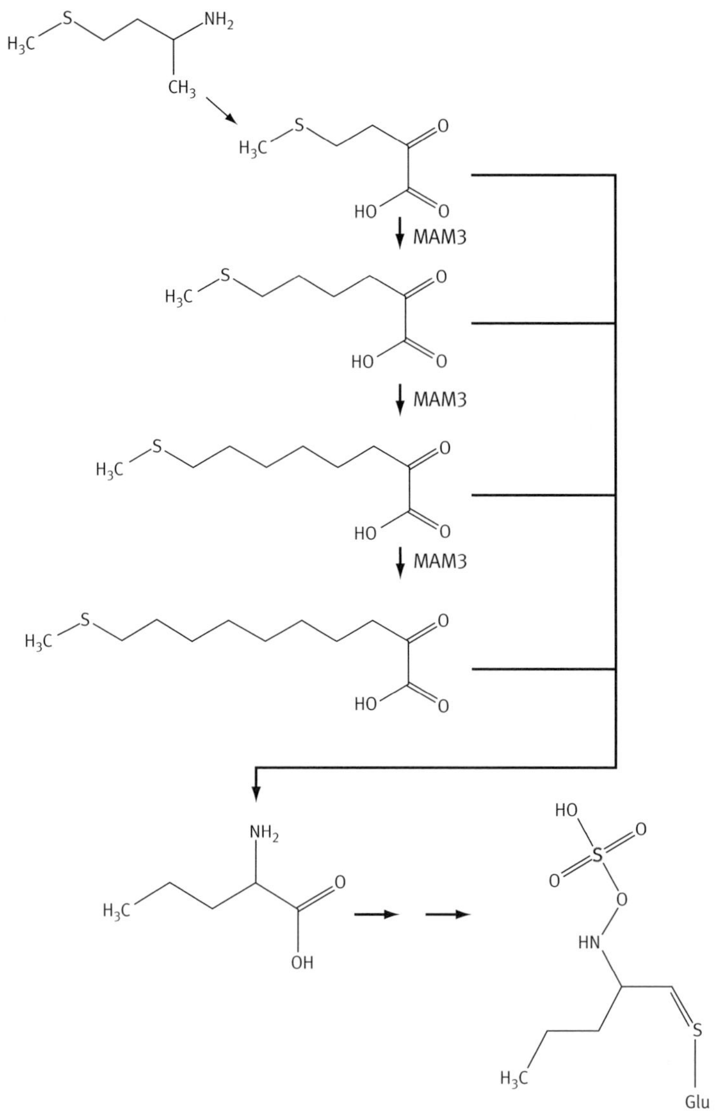

Figure 3.11. Diversity generation in the glucosinolate pathway.

Aromatic glucosinolates can also be made from tryptophan and phenylalanine. Adding to this first level of diversity, generated by feeding different amino acids into the pathway, the carbon chain length of some of the amino acids can be extended by an iterative two-carbon addition to give a variable chain length, each of which can enter the next phase of the pathway[19] (Figure 3.11).

The peptides

Peptides are short chains of amino acids. By combining in various ways, two or more of the over 20 amino acids that are made by all cells, a very large amount of chemical diversity can be generated. Some species of simple organisms such as the bacteria and the cyanobacteria make a number of distinct peptides, each species producing a different mixture. Some of these molecules have been shown to possess biological activity in certain test systems; hence these peptides could be considered as NPs. Interestingly, the biosynthesis of these peptides has been compared, as has the biosynthesis of other NPs (see Chapter 5), with nature's version of the chemist's attempts to maximise the production of chemical diversity.[20]

A few of these peptides have been shown to be some of the toxins associated with 'red tides', the population explosions of cyanobacteria that sometimes poison fish and other creatures in regions of the ocean (or indeed in inland waters). Because of the concerns about the potential contamination by such poisons of popular sea foods, such as shell fish, and the possibility that drinking water supplies could be tainted, these substances have been much more intensively studied in the past two decades. The full extent of the synthesis of peptides by organisms, the amount of diversity being produced and the evolutionary significance of these peptides are only gradually becoming apparent. Unlike all the other major groups of NPs, none of these peptides has ever played any significant part in human commerce; hence these compounds did not attract the interest of chemists and biologists until relatively recently.

What does this chapter tell us about the way science works?

Seeing the woods instead of the trees. There are hundreds of thousands of NPs, but maybe only tens of thousands of enzymes are involved in their synthesis and only a handful of basic strategies that have evolved to generate this chemical diversity. In this very cursory coverage of the biosynthesis of NPs, the detailed biochemistry has been omitted in the hope that the aspects of NP synthesis that are common to the major pathways can be seen more clearly. It is surely striking that evolution has apparently come up with very similar basic tricks to generate low-cost chemical diversity in each of the major NP pathways. It is only by looking at NPs as a large group that these similar strategies become evident. So, once again, the problem of the fragmentation of the subject, and the excessive concentration on detail, has made it harder for non-NP specialists to see the important feature (the wood) because their attention was directed towards the detail (the trees).

4

Are NPs Different from Synthetic Chemicals?

Every science has for its basis a system of principles as fixed and unalterable as those by which the universe is regulated and governed. Man cannot make principles; he can only discover them.

—Thomas Paine

Summary

Ever since the eighteenth century Swedish chemist Berzelius postulated that naturally made chemicals were different from man-made chemicals, that idea has lingered on, especially in the minds of the general public. Consequently, there is a great commercial advantage in claiming that a product is 'natural'. Many consumers perceive natural chemicals as being safer than man-made, synthetic chemicals; hence the increasing popularity of food grown without the use of synthetic fertilisers or pesticides. The key difference between synthetic and naturally made chemicals is that the former are usually made with the aid of harsh chemical reagents and the latter are made by enzymes. The same substance made by either method is identical, but the spectrums of substances made by chemists and by organisms differ a little because humans cannot at present economically make some structures using chemical reagents. Crucially, the method of making a substance does not predict all the properties of the substance made. Chemicals made by organisms are not inherently 'safer' than synthetic chemicals.

Are natural chemicals different from synthetic chemicals?

Among the general public, there seems to be a common belief that naturally made chemicals are somehow different, maybe safer, than synthetic chemicals. Even some well-respected chemists once believed that NPs were in some way different from synthetic chemicals (see Chapter 1) and there are still some advocates of that view among those seeking new pharmaceuticals. So what is the answer to the simple question posed at the beginning of this chapter? Unfortunately, there is no simple answer. The difficulty in answering the question as posed is that the question is ambiguous but explaining that ambiguity is a good place to begin to construct an answer. The underlying ambiguity to the question is whether one is seeking an answer about one specific chemical or the whole class of chemicals. If one asks whether a single named molecule differs in its properties depending on how it has been made, the answer is no. But if the question is asked about a group of chemicals the answer is more complex.

Is vanillin in *vanilla extract* the same as vanillin in *vanilla flavouring*?

Vanilla is a very popular food flavouring commonly found in ice cream or cream soda drinks and is not uncommon in foods and cosmetics. The pod of an orchid native to Mexico was used by pre-Columbian Mesoamerican people to produce the flavour and it was introduced into Europe in the sixteenth century. In 1874, two German scientists, Ferdinand Tiemann and Wilhelm Haarmann, proposed a structure for the chemical that was predominantly responsible for the flavour and named this compound vanillin. They also devised a synthesis and indeed started a company to make an artificial vanilla. In the twentieth century, new methods of synthesis and manufacture were discovered and synthetic vanilla became competitive with the extracts of the vanilla pod. There is no evidence that a vanillin molecule made by the vanilla plant (currently the source of about 1800 tonnes, costing several hundred $ per kg) can be distinguished from a vanillin molecule made in a chemical factory (>12,000 tonnes, priced at $15 per kg). However, the willingness of the consumer to pay for what they think is a natural product is evident by the fact that a company which developed a microbial fermentation process to make vanillin from ferulic acid can label its product as a 'natural flavouring' (fetching $700 per kg) despite the fact that it is produced in a factory. Vanillin is vanillin, whether made by a human or by any other organism (Figure 4.1). However, that does not mean that the *synthetic vanilla flavouring* and the *natural vanilla extract*, both widely available in many grocery stores, are identical. Both will contain vanillin as their main component but the natural extract is likely to contain some other NPs in very small amounts. The synthetic vanilla is likely to be purer, despite the fact that the natural extract is likely to be 100% more expensive. Clearly, many consumers are willing to spend much more to obtain the natural extract and that choice may be due to the consumer having a feeling that the 'natural chemical' is preferable but the preference might also be based on a desire to support Third World producers or simply a feeling that the best always costs more.

OK so an individual chemical is the same irrespective of the route of synthesis but are synthetic chemicals as a group different from NPs as a group?

A more complex analysis is needed when one asks whether naturally produced chemicals as a group are different from synthetic chemicals.

The first problem in answering this question is that there is a considerable human bias in selecting which synthetic chemicals have been made and which NPs have been isolated, purified and structurally determined. In other words, humans do not have a properly randomised collection of synthetic chemicals or randomly isolated NPs to do

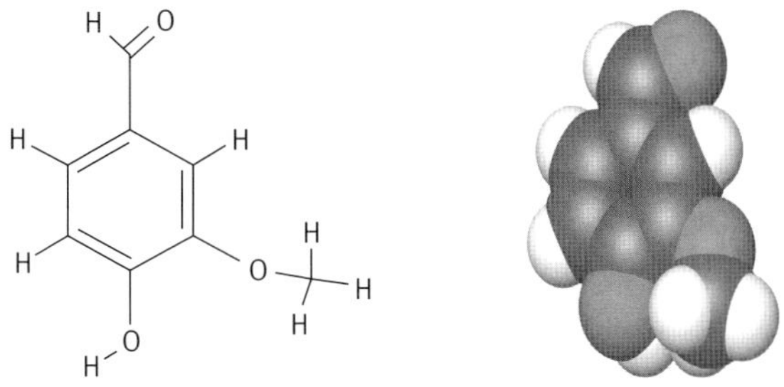

Vanillin (4-hydroxy-3-methoxybenzaldehyde)

Figure 4.1. The 2-D (left) and 3-D (right) structures of synthetic or naturally occurring vanillin, the chemical that gives vanilla extract its characteristic flavour and smell. Neither representation of the structure provides a guide to the fact that humans like its flavour or smell, or indeed that the human nose can specifically detect the molecule.

any comparison. Consequently, any collection of synthetic chemicals or any collection of NPs will inevitably possess some bias in its average properties simply because the selection for certain properties would have been instrumental in the process of adding each individual chemical to the collection. So, if a difference is found in the average properties of a collection of synthetic or naturally made chemicals, one has to be cautious about ascribing significance to the difference.

Systematic comparisons of the properties of collections of natural or synthetic chemicals were rare until relatively recently when the pharmaceutical industry began to re-evaluate the trend towards using only synthetic chemistry to generate new leads for novel pharmaceuticals. During the past two decades of the twentieth century, the cost of testing any chemical for its ability to influence the biochemistry of a cell fell very dramatically when robotic systems were introduced into the laboratory (see also Chapter 7). It became technically possible for a company to assess hundreds of thousands of chemicals a day to make a quick assessment of whether any one chemical possessed a specific biological property (e.g., to seek among a 'library' of chemicals an efficient inhibitor of a specific enzyme). Once such robotic methods became economic, the pharmaceutical companies needed more new chemicals to test. Most pharmaceutical companies already had 'libraries' of chemicals containing hundreds of thousands of different synthetic chemicals. Specialist companies also exist which will sell anyone a library of chemicals containing tens of thousands of synthetic chemicals. It was accepted that the more chemicals that could be 'screened' for pharmaceutical use the greater the chance of finding a winner. The pressure to increase the number of chemicals available in any one company led chemists to devise new approaches to making even more new chemicals for testing. For example, one new approach concentrated on making as much chemical diversity as possible without worrying too much about the purity of the chemicals being made. Traditionally, chemists devised methods to make a single target structure with great purity, but now chemists devised ways of producing mixtures of chemicals, some of which were unplanned. The cost of making a novel synthetic chemical was now well below the cost of isolating and purifying a novel NP from an organism. In the 1990s, nearly all the major pharmaceutical companies, who had been the main funders of chemists seeking new NPs, lost interest in NPs.[1] Consequently, there was a decline in the importance of collecting and characterising NPs as a means of finding new drugs (see also Chapters 2 and 7).

However, the ability to make and test hundreds of thousands of new chemicals in new cell-free systems proved to be less successful than expected in producing new pharmaceutical leads. It was not uncommon for a company to screen a library of tens of thousands of chemicals to find only a very few promising 'leads' (this fact is very significant for reasons that is explained in the next chapter). Soon the NP enthusiasts began to argue that the pharmaceutical industry was beginning to pay the price of neglecting NPs. Those advocating a return to NP isolation and screening pointed out that despite the increasing number of synthetic chemicals screened for useful pharmaceutical properties, pharmaceutical products based on NPs had been vitally important to the industry. These arguments gained theoretical support from the widely accepted view of ecologists that natural selection in organisms making NPs had selected for the retention of biologically active NPs.[2] It was argued by some NP advocates that, in essence, organisms making NPs were doing a 'pre-screening' process on behalf of the pharmaceutical companies because most NPs must have some effect on organisms because that is the reason why they are made (see Chapter 5 to understand why this argument no longer holds water). Maybe after hundreds of millions of years of evolution, organisms making NPs had evolved ways of making chemicals with structures that had a very high chance of being biologically active? Maybe these structural features were more common in collections of NPs and were unusual in collections of synthetic chemicals? If one could identify these crucial structural features maybe chemists could design synthetic chemicals which have a higher chance of possessing potent biological activity?

How does one make the comparison? Although one learns chemistry by looking at chemical structures as they are written on paper, this is unlikely to be very useful because biological activity is determined by the three-dimensional (3-D) structure of molecules. Even if one produces computer models of 3-D chemical structures, which features of the beautiful shapes does one concentrate on (Figure 4.1)? In the same way, every organism looks different to us because the organism's body has its own distinctive size, shape (morphology) and colour, so every molecule has its own 'morphology'.[3] Every molecule is made up of different atoms, and the same collection of atoms can be joined together in different ways. Molecules, like many organisms, flex parts of their structure; consequently, exactly what space they occupy might be quite hard to define, especially if their morphology can be influenced by the local environment. Those unfamiliar with thinking about the shape or properties of molecules might find it easier to use the analogy of trying to compare animals on the basis of the 3-D space they occupy. Think of the shape of a horse when it is lying down, bending to drink, standing in a

Table 4.1. A systematic comparison of the group properties of a collection of synthetic chemicals and a collection of naturally occurring chemicals, both collections held by a pharmaceutical company.

Parameter	Synthetic drugs	NP drug leads
Number of different molecules in collection	5,757	10,495
Average molecular weight	356	360
LogP_c (estimate of fat solubility)	2.1	2.9
Hydrogen donors per molecule	2.5	1.8
Nitrogen atoms per molecule	2.3	1.4
Oxygen atoms per molecule	4.1	4.3
'Rule-of-5' alerts (%)	10	12

Source: (Man-Ling Lee and Schneider, 2001)

pose, trotting, jumping and running—how could one reasonably produce a measure of the 3-D space a horse occupies? Of course, one can easily compare the properties of individual elements such as the length of the legs, ears, tails, the size of the eyes, weight, and so forth, but is there a measure of the 'horseness'? Despite these difficulties, comparisons have been made of the properties of collections of synthetic chemicals and collections of NPs. One such analysis by Man-Ling Lee and G Schneider[4] used sophisticated software tools to predict the molecular properties from simple structural data (Table 4.1). The average mass (molecular weight) of the NP and the synthetic molecules in the two collections were very similar as was the average fat solubility and the number of oxygen atoms per molecule. One significant difference was that the NP collection had a greater diversity of ring systems (1748) than the synthetic drug collection (807). There was an overlap of the type of ring structures between the groups, with the NP collection including 35% of the rings found in the synthetic drugs but only 17% of the NP ring structures were found in the synthetic drug collection. Some of the ring structures, unique to each group, are shown in Figure 4.2. Clearly, humans and other organisms can make quite complex ring structures but there are differences. Organisms can more easily construct molecules with atoms joined in rings. Why have these differences been found and what is the significance of these differences?

The main reason for the difference—reagents versus enzymes as synthetic tools

The collections of chemicals studied by Man-Ling Lee and Schneider were not truly random because they were seeking to compare collections that were considered to have a high chance of being useful as pharmaceutical agents. So there would be some positive selection for some properties that enhance the chance of a molecule having value as a drug. Hidden beneath the selection of the chemicals for comparison, there may well have been other negative selection factors—maybe highly toxic chemicals were

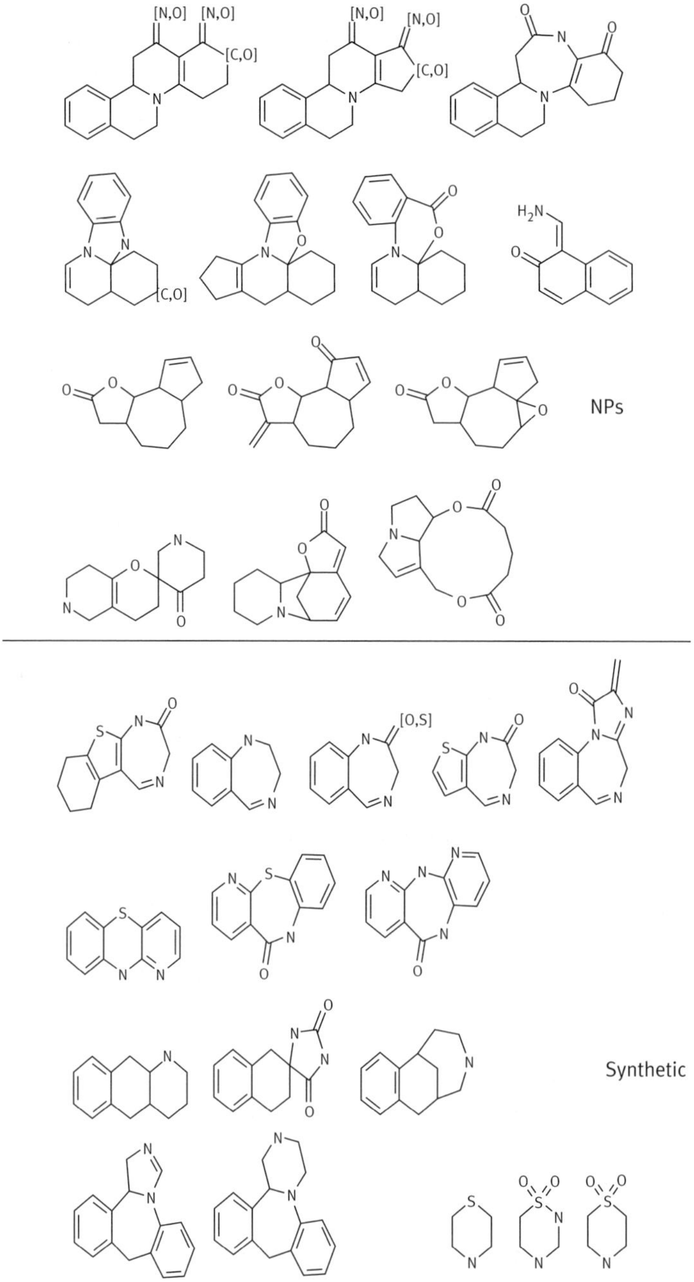

Figure 4.2. The ring structures are more commonly found in two collections of NPs and in synthetic chemicals (from Man-Ling Lee and Schneider, 2001). Even the untutored eye can spot that there are some differences between the two collections but what is the cause or significance of the differences?

excluded from both categories, maybe very rare chemicals were excluded from the collections of natural chemicals and maybe very hard to make chemicals were excluded from the synthetic collections. However, even given these possible caveats, one might be impressed by the fact that *there was no way of categorising a chemical as synthetic or natural on the basis of any of the properties that were assessed.* Yet there were a few differences, such as the difference in ring structures, so maybe such ring structures should be the focus of attention of those seeking to make new pharmaceutical products? Such a conclusion would actually be premature because there is one difference between synthetic and natural chemicals that could fully explain the apparent non-overlap of structures in the two collections. It is a difference that is so obvious and fundamental that it is easy to overlook! Organisms use *enzymes* to make NPs. Chemists use *reactive chemical reagents* to make synthetic chemicals. These two methods are not equivalent; hence, the spectrum of chemicals made using these fundamentally different synthetic methods will inevitably only partly overlap.

Chemical reagents versus enzymes

History shows that humans, at first gradually but lately more rapidly, have found ways of making increasingly complex objects, structures and devices. Human imagination and dexterity seems to have changed little during recorded history, so why has the rate of development of complexity of man-made devices changed. The answer seems to be that the complexity of the tools available to humans has increased. Simple tools can make simple objects with little skill but to make complex objects with simple tools requires very great skill. A hammer takes little imagination to make or to use but, used with imagination and skill, beautiful or practical objects can be made by a blacksmith. However, in the late eighteenth century and early nineteenth century, imagination was directed towards making more complex tools, tools that replaced the skill of the blacksmith with the precision of the tool itself—'machine tools' were invented. The development of machine tools allowed metal objects to be made economically, precisely and repeatably. Very sophisticated tools used by semi-skilled workers could produce huge numbers of identical objects, such as nuts and bolts. Very complex devices could be constructed by assembling in sequence the simple individual components made by machine tools. Within 100 years, highly complex machines like steam engines, locomotives, cars or aircraft engines could be mass produced. A blacksmith could make a modern car but huge effort would be required because the blacksmith's tools are simply not precise enough, even when used skilfully. What has this to do with chemical synthesis? The analogy being built is that chemical reagents are the equivalent of blacksmith tools and chemists can be regarded as very highly skilled chemical blacksmiths. Enzymes are the biological equivalent of machine tools. Each type of enzyme is a precision tool which does its specific task very well but it is not versatile. The complexity of naturally made chemicals can be achieved by combining many precise steps in a definite sequence.

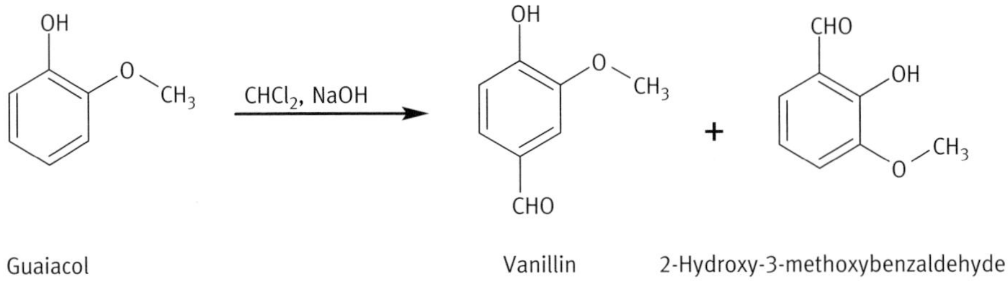

Figure 4.3. Using the example of one synthetic route to vanillin (see Figure 4.1), one problem of the lack of specificity of chemical reagents is revealed—two products, not one, are produced. If a long sequence of chemical reactions is needed to convert a cheap, readily available starting material into a valuable substance, the percentage of the desired intermediate gained at each stage is one of the major factors in judging the economic viability of the synthesis (another major factor is the cost of ridding the final desired product of the inevitable impurities produced by the side reactions).

Chemical syntheses

The clever chemist uses chemical reactions, usually brought about by the addition of reactive chemicals to a starting material, in a sequence to bring about a series of additions, rearrangements and deletions to the starting material or the intermediates made at each stage. With great skill, and sometimes with some luck, the desired final product is made. However, while this sequence of chemical transformations works well for simple, inherently stable molecules, the more complicated the molecule being made, the harder synthetic organic chemistry becomes. The problem is that the chemical reagents being used are only partially selective. Consequently, with the planned reaction taking place, there are very often 'side reactions' which are converting the precursor into unwanted products—the reagent is reacting on the 'wrong' part of the precursor chemical as well as the 'right' part. It is rare for the chemical reaction between two molecules to produce one product in 100% yield (Figure 4.3). In general, the more complex the structure being used as the starting material for a particular stage of the synthetic sequence, the greater the probability that the proportion of desired or undesired product will fall at each step of the synthesis. As there might be several steps, the overall yields declines and, eventually, the synthesis becomes uneconomic or impractical. There are numerous examples of very complex molecules being finally made by chemists, after years of effort and at great expense, but such huge efforts are only made when it is known that chemical might have huge value. For example, it has been reported that 1000 chemists sought ways of synthesising penicillin (produced naturally by the *Penicillium* fungus, see Chapters 2 and 7), yet they were unable to find a synthetic route to the antibiotic.[5] Meanwhile, it was soon discovered that the yields of penicillin obtained from fungal cultures could be greatly increased by strain selection and by optimising the culturing conditions used to grow the fungus. Consequently, attempts to synthesise penicillin

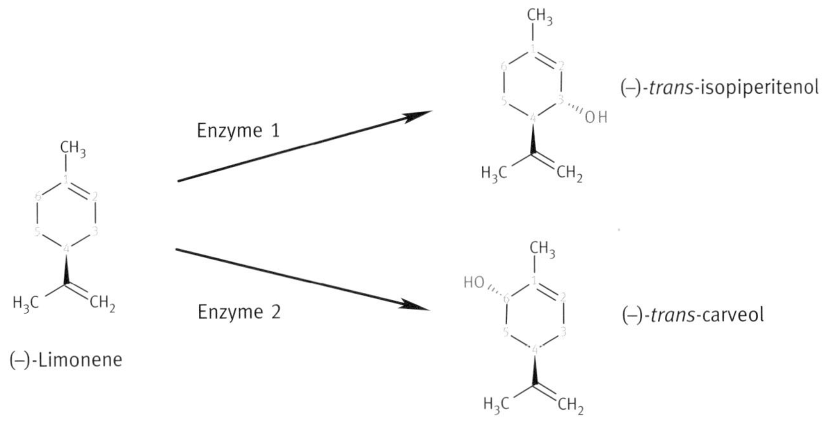

Figure 4.4. Enzymes can bring about a specific chemical transformation of a specific individual atom within a molecule. This is illustrated by using an example of the synthesis of two of the monoterpene intermediates leading to the characteristic scent of peppermint or of spearmint. *Enzyme 1* (cytochrome P450 (–)-limonene-3-hydroxylase), found in peppermint, introduces a hydroxyl group at position 3 of the ring. *Enzyme 2* (cytochrome P450 (–)-limonene-6-hydroxylase), found in spearmint, brings about the same type of chemical transformation but at position 6 of the ring. (These enzymes are chosen because they are part of the unfolding story in Chapter 5. Many other examples could be used.) Contrast this figure with Figure 4.3.

for commercial production were abandoned in favour of fermentation. Two other very important pharmaceutical drugs, the anticancer drug taxol (which is discussed in detail in Chapter 7) and the antimalarial drug artemisin, are further examples of chemicals which still cannot be made by humans economically; hence, supplies are still obtained from plant material.

Enzymatic syntheses

In contrast to chemical reagents, many enzymes can be impressively specific in terms of the kind of reaction that they can carry out and some enzymes will only act on very specific structures. This precision is called 'enzyme specificity'. However, as explored in Chapters 5 and 9, because of the fragmentation of the subject of biochemistry, outlined in Chapter 1, there has been widespread misunderstanding about the 'specificity' of enzymes. In elementary teaching of the subject, it is unfortunately implied that every enzyme can act only on a single substrate to produce a single product (Figure 4.4). This is now known to be a gross oversimplification which arose largely because early biochemists working on enzymes chose to work on enzymes contributing to basic metabolism where evolution has indeed selected such highly specific enzymes. It is now clear that some enzymes involved in NP biosynthesis are capable of acting on more than one substrate and a few can even produce multiple products (Chapters 5 and 9). However, even these less specific enzymes involved in NP synthesis are precision tools compared to chemical reagents; they act on specific parts of a molecule, bringing about a precise change to one part of the chemical structure, with no effects elsewhere in the structure. By this means, the 'delicate' naturally made structures that so impressed Berzelius (Chapter 1) can be made, structures that are very hard to reproduce efficiently with any sequence of chemical reagents.

The consequences of the synthetic differences

For the reasons outlined, it is clear that any collection of synthetic chemicals will inevitably be constrained by the fact that only certain structures are possible within the budget devoted to the programme. There will be a predominance of easily made structures, which are the end results of limited reagent repertoires. Because chemical reagents are sometimes quite harsh, particularly reagents used on a large scale industrially, it is predictable that synthetic chemicals might on average be more chemically stable than the average NP because synthetic chemicals have to survive the harsh processes used to make them. Likewise, any collection of NPs will also be non-random because human choices will have been made at all stages of the isolation and characterisation programmes. The two most decisive features in choosing which organism to isolate NPs from, and which molecules to characterise from those organisms, are the known biological activity of the material and the quantity of material that can be easily obtained. The plants in human cultivation, or which humans had particular knowledge of, other than food or material crop plants, were heavily biased towards species that had a supposed biological activity (as evidenced by their use as medicinal plants or herbal remedies). The fact that morphine was subject to intense chemical study even in the early nineteenth century was not because it was chemically fascinating but because it had such potent biological activity, a property of great value to humans. Thus, collections of NPs will inevitably be drawn from a subset of available species on the basis of some human judgement as to the importance of that species to human affairs. Once any species is chosen for analysis, the analyst is immediately faced with the problem that he/she will encounter thousands of different molecules as he/she begins to probe the composition. Some molecules will be present in large amounts and some in minute amounts (Figure 4.5). Some substances will have some unique molecular properties that make them relatively easy to purify, others will hide in a background of similar but different molecules. Some chemicals will be very stable and hence easy to work with; others will be destroyed at the early stages of analysis. It is not surprising that any collection of NPs will tend to favour stable compounds that occur in relatively high concentrations and which have some molecular properties that enable their isolation and purification to be carried out economically.

In summary, it is predictable that collections of synthetic chemicals and of NPs will not be identical because their methods of synthesis are fundamentally different. However, the differences might be of little predictive value when seeking chemicals with specific properties or functions.

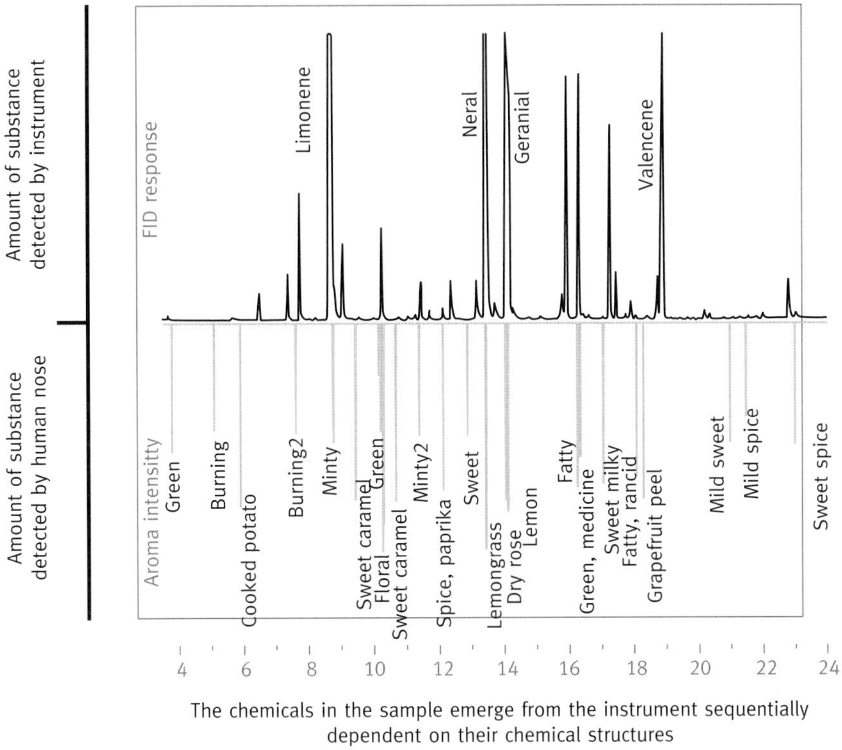

Figure 4.5. The NP composition of a typical plant extract (in this case, a citrus hybrid) as revealed by gas chromatographic analysis. The peaks on the upper trace represent the different chemicals detected by the instrument, with the peak area being a measure of the amount of any substance. Note that there are a few major NP peaks but even more very minor ones. The spikes pointing down on the lower trace are the odours detected by a human 'sniffer' with their perceived odour name. Note the human detection of odour does not always correspond to the emergence of a major chemical peak. For example near the start of the analysis, the 'green' or 'burning' smell detected by the human does not correspond to any instrument detection so those chemicals are below the level of detection of the instrument. (Modified from the data of Morton M, Smoot JM, Mahattanatawee K, Grosser J and Rouseff RL, Citrus Research and Education Center, University of Florida.)

What does this chapter tell us about the way science works?

In 2008, the top 10 pharmaceutical companies spent \$50 billion on research and development (see Figure 2.1 for a NP reference point). Sadly, much of the research data they accumulated will never be made publicly available, so it is not easy to analyse the results of cross-industry research. It is possible that the major pharmaceutical companies possess data which would allow a much better analysis of the possible differences between collections of NPs and collections of synthetic chemicals. Even gaining good data about the probability of any one chemical showing potent, specific biological activity is extremely difficult, yet the information does exist in the computers of many pharmaceutical companies. One wonders whether the public interest would not be better served if all research data were to be made freely available, may be after 12 months in the case of academic research and possibly after 10 years in the case of research data held by commercial companies. It is quite remarkable that the issue of who 'owns' data is still very uncertain. For example, in the United Kingdom, it is extremely vague as to who 'owns' the data produced by government funded research in universities (see Chapter 7 for examples of arguments about the patenting of some NPs). Should research data be 'owned' by the grant holder (despite the fact that most data are now obtained by teams of technicians, graduate students and postdocs) or the institution employing the researcher(s) or the organisation(s) that funded the work? The past 25 years tipped the balance towards giving 'ownership' to individuals, by encouraging them to patent discoveries in the expectation that 'wealth' would be created. It is now widely accepted that this commercialisation of scientific research hinders the exchange of ideas and data, making universities and research institutes behave more like companies who guard their data with the diligence of Cerberus.

5

Why Do Organisms Make NPs?

In some scientific circles it is something of a sport to theorize about function, often with the intent of finding one overriding axiom true for all secondary metabolism. Speculations range from the notion that they are waste products or laboratory artefacts, to the concept that they are neutral participants in an evolutionary game, to ideas of chemical weaponry and signalling.

—*Bennett, 1995*

Summary

There have been many attempts to explain why organisms make NPs. The most widely accepted model, the Chemical Co-evolution Model, proposed that the interactions between a plant species and insects (which interact positively or negatively with the plant) were shaped by NPs made by the plant. For example, a plant making a novel NP would gain fitness if that NP reduced the fitness of herbivores or the plant would gain fitness if the NP had a beneficial effect on insects whose visits benefited the plant. The evolutionary response of the insects would be to adapt to these new selection pressures and this would result in the insects becoming increasingly specialist. This model argued that the great diversity of NP structures resulted from the great diversity of the co-evolutionary processes in the natural world. The model was easily applied to plant–fungal interactions but was least convincing when explaining the more ancient NP diversity in microbes.

The Chemical Co-evolution Model was based on the assumption that every NP possessed (or had possessed at some stage in evolution) some biological activity that enhanced the fitness of the producer. This assumption is not supported by experimental evidence and the assumption has no theoretical basis. Extensive studies of collections of both synthetic chemicals and NPs have shown that the probability of any one chemical structure possessing potent, specific biological activity is very low. These experimental findings are supported by the current understanding of the way in which small molecules interact with proteins to bring about biomolecular activity.

The Screening Hypothesis seeks to explain the evolution of NPs when the chances of any one NP benefiting the producer are indeed very low. This simple hypothesis predicts that certain metabolic traits which favour the generation and retention of NP diversity will be retained during the evolution of NP metabolism. The most important of these predicted metabolic traits is the ability of enzymes making NPs to either accept more than one substrate or to make more than one product. Evidence consistent with the Screening Hypothesis has grown in the past decade such that it is now a credible model based on sound physicochemical and evolutionary principles.

Some criticisms of the Screening Hypothesis helped refine the model and the thinking behind the model has since been applied to a wider range of metabolism (see Chapter 9).

Reductionism—a scientific tool only as good as its user

Our understanding of the natural world comes from identifying, classifying and determining the function of its components. The process of gaining information about the natural world has progressed at a rate dependent partly on what was technically possible to those carrying out the studies and partly on the state of intellectual development. One of the most successful approaches to carrying out scientific studies has been reductionism. The investigators attempt to identify the key components that make up a large, complex process. They then study the individual parts in isolation, trying to understand what each part does. Once the role of every part is established, a view of the overall process can be formed by assembling all the individual ideas in a coherent, logical manner. A good analogy of the benefit of studying such complex objects in this way would be the study of a car. Most people can grasp the concept of what a car is capable of doing but trying to understand how a car works requires the attention that is paid not to the whole object but to the individual parts. The whole object rolls back or forward because it has wheels. The object can move in different directions because the front wheels pivot. The force to turn the driving wheels comes via gears and shafts from another complex item that will function in isolation—the engine. However, when using a reductionist approach, caution is needed in choosing the appropriate level of scale for the analysis. Taking a car completely apart, to the final screw and spring, without understanding the purpose of the component from which the screw or spring is being removed starts to reduce the quality of information being accumulated. One would have little chance of understanding what a car was if one only had a huge heap of individual parts not organised in any way. So, reductionism is a powerful tool, but only if used wisely.

Using the car analogy to illustrate the power and problems of reductionism should not encourage the view that biological systems are simply like machines. One key difference is that 'the evolution' of the car and the evolution of an organism are very different processes. Humans in their endeavours can be informed by experience and knowledge but at any time humans can introduce radical changes—a new design can start with a clean sheet of paper and a new design can be radically different from anything made before (e.g., front wheel drive vs. rear wheel drive). In human artefacts, there is 'an evolution' of thought in the design of the object but the manufactured object can be unrelated to any previously made object. In contrast, biological evolution always starts with an existing system, then many minor variants are made which are based entirely on the original design and each variant is allowed to compete against the original and all other variants. If a variant is made that performs better than its competitors then it will have a better chance of reproducing and its genes, which code for the new variant, will increase in frequency in the population.

Reductionism has been very evident in studies of NPs. The study of NPs passed quickly from herbalism to the study of chemicals at a time when evolutionary thinking was only just developing. Because those chemical studies were taking place in Chemistry Departments (see Chapter 1), the reductionist approach tended to concentrate on characterisation of yet more and more chemicals. Biochemists, universally more schooled in chemistry than evolutionary theory, tended to focus on the role of their favourite enzymes in making one type of NP—the normal reductionist approach adopted by biochemists. When biologists began to participate in exploring NPs, each biologist tended to study one particular NP (or maybe a related family of NPs) and very often they concentrated their thoughts on one process in their chosen model organism. Thus, those working on NPs were not only fragmented in different scientific disciplines but also each had, rather too rapidly, reached a very high level of reductionism (the analogy of the car being stripped to individual components before the main functional parts were understood comes to mind). While the reductionist approach was successful in each area, too few people stopped to ask the question as to *why* organisms had an ability to produce all these wonderful, striking chemicals. By the middle of the twentieth century, there was no shortage of information about the tens of thousands of individual chemical structures (published in specialised chemistry journals), about the biochemical pathways that form each major class of NP (published in biochemistry journals), about the properties of some enzymes in these pathways (sometimes published in even more specialised biochemistry journals) and about the properties of some NPs that had notable effects (often published in specialised medical journals or cell biology journals). Towards the end of the twentieth century, this fragmentation actually increased as more biologists took an interest in NPs but they tended to publish their work in even more specialist journals, journals often targeted at even smaller groups.[1] However, in this chapter, we shall leave the wealth of detail in the background and try to show how the few people thinking about *why* organisms make NPs were developing their ideas.

The development of ideas about why organisms evolved to make NPs

Nineteenth-century scientific work on NPs

From the fifteenth century onwards, the major European powers were sending plant collectors to scour the world for new, exotic plants and the number of species that were accessible to herbalists and physicians increased. The great universities began to assemble plant collections in Botanical Gardens and some royal collections were also established, the best known perhaps being Kew Gardens in London. The reasons for gathering such collections varied. Some scientists desired to have access to a wide range of fascinating biological specimens in order to try to understand the natural world in all its diversity. Others wanted to ensure that they had access to specimens of all plants that may have commercial value so they could be introduced into new colonies (as outlined in Chapter 2, high value plants were often jealously guarded to maintain a monopoly of supply). However, some plant collectors had neither scientific nor commercial motives, rather they were simply obsessed by the possession of rare specimens (although as the Dutch *tulip mania* had demonstrated in the seventeenth century, rarity can get translated into commercial gains for some but ruin for others).

Of course, there was a widely held view in nineteenth-century Christian Europe that God had created all organisms for man's benefit (it was *man's* benefit), hence many considered that the existence of NPs as requiring no explanation. However, some scientists began to see the need to seek explanations of how the natural world had been formed.[2] A common theme began to emerge—the natural world was not a fixed entity that had been created as it now existed, rather the world as we experience it at any time is changing due to the action of various natural forces. Geologists, led by the Scot James Hutton, speculated how mountains arose, why great valleys had been formed and why soils varied. Biologists, building on the work of the Swede Carl Linneaus, had begun to understand the relationship between different organisms. After Wallace and Darwin had outlined the principle of natural selection, it was soon widely accepted that individual organisms were subject to competitive selection. An individual making something, which has a significant cost of manufacture or maintenance, that brought no increase in fitness would be less fit than an individual that lacks this redundancy. This principle would apply even at the chemical level. However, as introduced in Chapter 1, the study of NPs during Darwin's lifetime was largely a matter of determining the structure of individual NPs and trying to find ways of making them synthetically; this work was carried out almost exclusively by chemists who were largely uninterested in why organisms made these fascinating structures. That set a pattern that was to continue for a century, where the study of NPs as chemicals was divorced from the study of the organisms that made them.

Twentieth-century scientific work on NPs

Throughout much of the twentieth century, chemists continued accumulating information on the structure, and sometimes the synthesis, of individual NP chemicals. A massive literature grew about all the weird and wonderful chemicals. These wonderful structures challenged and fascinated generations of NP chemists. Even after biochemistry departments began to appear, the study of NPs largely remained in chemistry departments, because biochemistry department were heavily biased towards the study of the major anabolic and catabolic pathways found in most groups of major organisms. Journals that specialised in the chemistry of NPs were published long before journals appeared that asked why organisms made them.

Despite the general lack of curiosity of why organisms might be making these important molecules, a very few individuals in the first half of the twentieth century did begin to explore various ideas as to why some organisms (it was now clear that microbes as well as plants made NPs) made NPs. However, the diversity of NP structures made it hard for anyone to come up with a universal explanation. This lack of a universal explanation was not of great concern because the increasing importance of NPs as pharmaceutical products in the second half of the twentieth century (led by the discovery of penicillin as outlined in Chapter 7) gave NPs an importance beyond whatever role they played in the organisms that made them. The huge new economic importance of the antibiotic NPs stimulated a demand for NP chemists. The other academic discipline which gained from this new interest in NPs was microbiology. Microbiology departments had flourished in the early twentieth century because many common diseases were caused by microbial infection and the ability to identify microbial contamination of food and water offered the potential to improve public health. Although chemists were the first to offer chemicals to attack pathogen organisms, after the discovery of microbial antibiotics, the selectivity and potency of these agents gave microbiologists a new role in helping fight infectious diseases. However, interest in NPs in the more mainstream biology disciplines, such as botany or zoology, remained small. This only changed a little in the third quarter of the twentieth century but by the last quarter, the study of the role of NPs had become respectable for a wide range of biologists. This was helped by the fact that modern agriculture was fighting a constant battle with crop pests and diseases; attempts to understand the natural defences of plants, postulated to involve NPs, was timely.

As noted in the quotation at the beginning of this chapter, towards the end of the twentieth century, some scientists even played down the importance of asking why NPs were made by organisms and seemed unembarrassed by the failure of those studying NPs to have a convincing universal model. As is sometimes the case when no universal model for a phenomenon is widely accepted, the problem is not a lack of theories rather there are too many. The problem was that none of the theories seemed to explain all the NP diversity that chemists continue to find.

Several ideas were advanced for why organisms made NPs

Waste products

Animal physiologists had long worked out that animals had evolved sophisticated structures in order to deal with waste products. Thus, it somehow entered the thinking of some biologists that all organisms would have to have some equivalent mechanism to get rid of waste. Because plants and microbes clearly did not have livers or kidneys, but they did have all these weird chemicals, it was proposed that these two facts were related. Maybe plants and microbes had no choice but to make weird chemicals in order to get rid of some other chemicals that were troublesome for them? Consequently, NPs were just waste products. What is so remarkable with this idea is that it was ever taken seriously. It shows such a complete misunderstanding of plants and microbes.

Why do animals produce waste? Each animal species has evolved to be able to survive on its own unique diet. However, all the different animal diets have one thing in common—they are made up of a complex mixture of complex chemicals and in many cases the diet will change with time. Animals vary in their ability to cope with a varied diet. Some species have evolved to be specialists, living on a very limited range of hosts (e.g., aphids). However, many other animals are generalist. Generalists have to survive on whatever they can find in their vicinity and have to be adaptable. In all cases, however, seeking the perfect food source—one that contains exactly the elements needed, in the right proportion and nothing else—is a risky strategy so evolution has favoured animals either have a digestion system that is versatile enough to take nutrition from various food sources[3] or an excretion system that can rid the body of all the unwanted chemicals ingested or created by the digestive processes. Consequently, even though each species may have some ability to select its food intake, there will be an inevitable mismatch between the chemicals needed to sustain the animals and the chemicals in the diet. In order to ingest the chemicals needed for sustenance, some unwanted chemicals must be taken in by the animal. In the extreme case of a generalist, on some days there may be too much protein yet on another day there may be too much carbohydrate or fat. On many days some chemicals will be ingested that simply cannot be assimilated. Consequently, the digestive system of each species has evolved to remove what that species needs from its mixture of ingested chemicals and to excrete the remainder as waste. The fact that one species of animals (e.g., a dung beetle) may gain its nutrition from the waste of another species (a cow) illustrates the fact that the term 'waste product' is a relative term.[4]

Why should plants produce waste? Plants and microbes, unlike animals, do not ingest chemically complex materials in order to gain nutrients. In contrast to animals, plants take in a very few simple molecules (water, carbon dioxide, nitrate, phosphate, other ions, etc.) which they use to elaborate more complex molecules. The plants have evolved to control the uptake of many of these elements, including all those elements used to make NPs. The plant with access to surplus carbon (available in infinite amounts in the air as carbon dioxide) can reduce the rate of photosynthesis or can store any surplus carbon in starch for example. A plant that suddenly finds its roots exploring a nitrogen-rich part of the soil (maybe where a worm is decaying) need not be stressed by the now unbalanced supply of nitrogen relative to carbon because the roots could take up less nitrogen, or more likely the surplus nitrogen could be taken up by the plant and stored in a storage protein. Likewise, for nearly every element, the plant has some control of the rate of uptake and usually has a route to an appropriate storage compound. Clearly, plants, in complete contrast to animals, have evolved to absorb a simple, predictable mixture of chemicals and the concept of waste is largely inappropriate. Furthermore, when thinking about the possibility that NPs were waste products, why would each plant species have evolved a novel NP composition given that they and their ancestors would have taken in the same mixture of simple molecules. In other words, the proposal that NPs are waste products is based on a false premise (that plants produce waste) and cannot even explain NP diversity.

Why should microbes produce waste? Many microbes, like plants, have considerable control over their intake of simple substances. Although many microbes do use complex molecules as a source of necessary elements, microbes commonly excrete degradative enzymes that breakdown the complex molecules into simpler substances, substances that can be taken up by the microbe in a controlled way. A characteristic of many microbes is their ability to alter their immediate environment and that includes their chemical environment. However, because of the diversity of microbes and their remarkably versatile biochemistry, one can never dismiss the possibility that some NPs made by microbes could serve a minor role in 'waste management'.

NPs are test chemicals made by 'Inventive Metabolism'

Given that natural selection operates by selecting the fittest from all the variants in a population, a mechanism must exist to generate the variation needed. Mutations causing small changes in the base sequence in the genome are the main source of this variation. It is easy for biologists to see the result of this variation if the changed phenotype is evident at the level of the whole organism. Such morphological variation, for example, was evident to those selecting features in domesticated plants and animals for thousands of years. However, biochemical variation is less easy to perceive because, unless it is manifested by a large visible change (e.g., flower colour, leaf variegation or fruit colour), the changes are hard to detect. Furthermore, the concentration of individual chemicals sometimes changes continuously on a daily and seasonal basis and as such it is hard to evaluate the basis of some biochemical changes, even if one can measure them. Because of homeostatic controls that also operate to maintain biochemical concentrations within certain tolerable limits, a mutation causing a change in the capacity of the cell to make a certain chemical might only be evident under certain specific conditions. However, the selection of variants from populations of plants show that even for NPs, individuals differ, sometimes remarkably, in their NP composition. The variation can be of two kinds. The relative amounts of each of the many NPs being made by species may differ. For example, a mutant plant that smells different from all others might only have increased the concentration of an existing compound to a level that brings it into the range that the human nose can detect. This kind of variation is relatively simple to explain in terms of a mutation having an effect on a pre-existing control process. However, more challenging is to explain how mutations produce new chemical entities. Zähner, discussing how microbes might produce new chemical diversity, proposed that NPs were the route to new chemical structures. He proposed that the metabolism that generated NPs was retained by microbes in order to retain a capacity for what he called *biochemical inventiveness.*[5] This imaginative idea seems to have been a minority view but it freed any individual NP from having a particular role, assigning a role instead to the overall metabolic capacity. One major problem with the model in evolutionary terms was why so much inventiveness was retained in a population. It would normally be expected that any mutant that possessed a novel, but useless, biochemical capacity would be lost from the population because the cost of making useless chemicals would make that individual less fit. Thus, the inventiveness would be expected to exist more at the level of the individual rather than the population. Although individuals in a microbial population might be predicted by Zähner's model to possess different NP compositions, the model could not easily explain why many individuals in a population of one plant species would all possess such a rich and characteristic NP composition simply to retain this inventiveness. Another challenge to Zähner's hypothesis was that it did not adequately define for what purpose the biochemical inventiveness was retained. How would throwing up new chemical structures enhance the fitness of the producer? More specifically, how could any of the new structures generated by such biochemical inventiveness integrate into the main metabolic functions of the organism in a way that would enhance fitness? The structures of most NPs are very different from the chemicals that are universally made by most living organisms.

NPs are relics of previously important cell regulators

Another microbiologist, Julian Davies, was struck by the fact that some NPs found in microbes had powerful inhibitory effects on the synthesis of some very important pathways. He argued that some of the powerful antibiotics found in microbes were relics of regulatory molecules that had once been used by microbes to regulate their own biosynthetic activities.[6] The basis of this proposal was that some antibiotics are so potent and specific in their ability to inhibit very basic mechanisms in the other microbial cells that they must have, at some time, evolved to act on those specific inhibitory sites. The main problem with this model is that it cannot account for a very large number of NPs that have very low biological activity and no known antibiotic activity. Davies's theory is too focused on antibiotics to be a general explanation for the existence of all NPs. The fact that some antibiotics have these poisonous effects on some cells is actually a very biased piece of information because antibiotics have been selected by humans from thousands of NPs found in microbes because they possess *unusual* properties. Hence, these unusual properties are unlikely to offer much help in understanding why the microbes that make them also make many more chemicals with no great potency as antibiotics. At best, the model could account for only a very minute fraction of NP chemical diversity, hence is unattractive as a general model to explain NP diversity.

NPs are made for no reason—they are fortuitous

As the twentieth century progressed, a number of microbiologists found that all the explanations advanced for why microbes made NPs were unconvincing when applied to their microbial NPs. Hence, many simply accepted that microbes made all sorts of weird chemicals because the costs associated with their production were so low that their production was not selected against. Evolutionary biologists were always very sceptical about such arguments but not all microbiologists interested in NPs were troubled by the views of evolutionary biologists. Furthermore, a microbiologist interested in finding new antibiotics often felt that evolutionary arguments were not going to help them in their quest.

NPs serve many roles in influencing the species–species interactions—the Chemical Co-evolution Model

One of the biggest problem that scientists face when trying to devise a theory to explain some piece of complex biology is knowing which pieces of information they should use as the building blocks of their model. By the last quarter of the twentieth century, the amount of information published about NPs was vast. The chemical structures of tens of thousands of NPs had been reported; the biological properties of many of these had been studied (at least to the extent that they had been passed through some drug screening trial by a pharmaceutical company or a government agency); thousands of papers had been published about the enzymes or pathways that make NPs and there was an increasing interest in the way that organisms that made NPs varied their concentration after those organisms were challenged by other organisms (see Chapter 8). Scientists still struggling to find a convincing reason why organisms made NPs were particularly impressed by two pieces of general information from the great pile that was NPs research data:

- Some NPs had very powerful, specific types of effects on some organisms.
- Some organisms increased the rate of synthesis when they were attacked by other organisms.

Surely, this could not be chance? A paper by Fraenkel[7] was particularly influential among plant biologists. Fraenkel marshalled the arguments that NPs were part of a chemical defence system that enabled plants to defend themselves against herbivores (plant eaters). It was an elegant and timely argument, advanced to a receptive audience. Around this period, plant pathologists were trying to discover more about the way in which plants defended themselves against fungal pathogens. For decades, synthetic chemical fungicides had been known to be very effective at protecting plants from fungal attack and the possibility that plants contained their own antifungal substances was being investigated[8] (see Chapter 8). Horticulturalists and plant breeders had known for many years that some cultivars of a species would resist fungal attack while other cultivars were highly susceptible to infection. It was also known that closely related fungal pathogen isolates could vary greatly in their ability to infect their host species. Maybe highly pathogenic strains of a fungus arose from individuals that had evolved a resistance to the fungicide produced by the plant? Very shortly, after highly toxic chemicals were introduced as control agents in agriculture, evidence that organisms could develop resistance to toxic compounds had begun to accumulate.[9]

The evolution of resistance to a natural control agent in a pathogenic organism and the evolution of new chemical defence chemical in the now susceptible host could be viewed as an 'arms race' (a term all to familiar to most adults at the time when these ideas were blossoming). Fraenkel's ideas were taken a stage further in a classical paper by Paul Ehrlich and Peter Raven[10] in which it was argued that co-evolution between specialist insect herbivores and their host plants could explain why so many insect herbivores were such specialists. An insect herbivore species that had adapted to cope with the specific chemical defences of a particular plant species would have more exclusive use of that resource but this would impose a new selection pressure on the plant species. Consequently, a mutant in the population of that plant species which made a more effective mixture of NPs would suffer less herbivory; it would thrive and be more successful at passing its genes on to the next generation. However, this change in NP composition of the plant would impose a new selection pressure on its insect herbivore species and mutants in those insect species that were better adapted to the new mixture would be expected to appear and subsequently thrive. These cycles of adaptation between the insect and the plant are an example of *co-evolution*, a process that results in a closer and closer association of one species with another. The beauty of this model was that it seemed to explain why plants produced so many NPs. There were so many different interactions taking place between plants and organisms that attacked them, or were attracted to them, that it was predictable that the many different selection pressures unique to each interaction would result in a very large NP diversification. Furthermore, this co-evolution model would explain why different species of plant produced quite different mixtures of NPs—every interaction would be unique and would have produced outcomes over evolutionary time that were unique to that interaction. This Chemical Co-evolution Model quickly became the accepted paradigm among plant biologist, entomologists and plant pathologists. However, many researchers working on NPs in microbes were still unimpressed by this model. Partly, this was because much less was known about the interaction between microbes, especially in the natural environment. The idea of chemical warfare between microbes certainly seemed superficially attractive in explaining why penicillin had been discovered but microbiologist had already learned that finding powerful antibiotics was not as easy as the chemical arms race would suggest (see Chapter 7). Many microbiologists worked for companies that had invested large amounts of money seeking antibiotics and, unlike NPs researchers working on plant–plant interactions or insect–plant interactions, they had purposefully surveyed the occurrence of antibiotics in a huge range of microbes and they had found very few. However, gradually some of those working on microbes began to rally around the Chemical Co-evolution Model, even if many remained sceptical. Commenting on this continuing scepticism Demain[11] said:

> It has always amazed me that the importance of chemical compounds in ecological interactions between plant versus herbivore, insect versus insect, and plant versus plant has been universally accepted, but the importance of antimicrobials in microbial interactions has been almost universally denied.

Constitutive versus inducible chemicals

In evolutionary arguments, it is always useful to try to identify where 'costs' arise and where 'benefits' accrue. For those not familiar with biological systems, it may be easier to think about the concepts using simple manufacturing economics as an analogy. A producer of any product has a number of costs—the cost of materials needed to make the product, the cost of the machinery needed to make the product, the cost of the energy needed to operate that machinery, the cost of any control systems (machines or people) that optimise the production process, the cost of maintaining all parts of the factory, the cost of keeping any stock and so forth. The benefits in such cases are easier to list—the resource that the company obtains from the market and from its customers. In the perfect market place, if two companies make identical products, the company with the lowest costs will thrive and eventually drive its competitor out of business.

If an organism is making any chemical or structure to serve a particular purpose, it clearly pays to make it only when needed (see also Chapter 8). If one makes a defensive chemical continually, there is an ongoing cost but the benefits may only occur irregularly. A mutant that only makes the chemical when the need for defence is sensed will clearly have increased fitness—equal benefit but lower cost. This concept was very influential in the case of the chemical defence of plants to fungal attack. Indeed, the definition of phytoalexin was 'an antifungal chemical made by a plant in response to fungal attack'. The idea gained wider acceptance somewhat later in plant–insect interactions but it is now generally regarded as an important feature in judging whether a particular chemical is involved in a particular interaction. The fact that an organism makes a particular chemical in response to a specific challenge has been taken to imply a powerful connection that indicates purpose (as we shall see later in Chapters 8 this logic is maybe not as good as it seems). The concept of the inducibility of NPs so nicely complemented the Chemical Co-evolution Model that both were strengthened by the mutual support.

The problem at the heart of the Chemical Co-evolution Model

As evidenced by the quotation at the beginning of this chapter,[12] by the end of the twentieth century, there was still no winner in the race to provide a universal model to explain the evolution of NP chemical diversity but the Chemical Co-evolution Model was well ahead. Indeed, many of its advocates considered the race won. Certainly, this model had many attractions but it had some worrying weaknesses.

The first weakness was that the Chemical Co-evolution Model was least convincing when applied to the most diverse, most ancient NP producers—the microbes. As explained above, many microbiologists simply did not find the model convincing when applied to microbes because very little real evidence for the model came from studies of microbes. The main evidence arising from studies of microbes was that some (a very, very few) microbial species could make antibiotics, but some microbiologists were much more impressed by the fact that the majority of microbes did not seem to make antibiotics.

A second major weakness for the Chemical Co-evolution Model was that it demanded that every NP has (or had in the evolutionary past) a specific role—every NP should possess an identifiable type of *biological activity*. Why should this be true? Because in simple evolutionary terms, mutants of an NP-producing organism that synthesised a new NP that possessed beneficial biologically active would be favoured, while a mutant that produced a new NP that was biologically inactive would not be fitter, but because of the extra production costs would be less fit; consequently, that mutant would be lost from the population by selection. This simplified version of the evolutionary theory to explain NP production would predict that any one NP-producing organism should produce very few, highly biologically active NPs.[13] Yet when one surveys the literature concerning the incidence of biological activity in collections of NPs, the percentage that have potent, specific biological activity is very, very low (usually $<1\%$). For example, after the discovery of penicillin, huge, well-resourced searches were conducted by many groups hoping to find yet more antibiotics. A 10-year study of 400,000 different microbial cultures only found three utilisable antibiotics. Another one-year study of 21,830 isolates found two possible substances with potential as antibiotics.[14] Once again one can speculate that there might be many more antibiotics being made by these cultures that were simply missed in the screening methodology but no model should confidently rely on speculation. So, the central problem for the Chemical Co-evolution Model is that it predicts that a very high proportion of NPs will have very potent, specific biological effects, yet the experimental evidence does not support this prediction.

Building a new model to explain NP diversity—the Screening Hypothesis

Instead of trying to sustain the Chemical Co-evolution Model by disregarding evidence which shows that any one chemical structure has a very low probability of possessing a specific, potent type of biological activity, what happens if one tries to build a model of NP diversity which accepts this fact? The result is the *Screening Hypothesis*, which takes the Chemical Co-evolution Model as its starting point but rebuilds it on proper physicochemical principles.[15] However, a sizeable digression is needed first to convince the reader that potent, specific biological activity is indeed a rare property for any one chemical.

What is biological activity?

The widely used term *biological activity* is so vague that it is virtually meaningless.[16] Even though the term *biological activity* is one that is widely used, and understood by most biologists or biochemists, it has no precise meaning without a reference point. As the great mediaeval scientist and mystic Paracelsus (1493–1541) noted, one cannot make statements about a form of biological activity without taking into account the concentration of the chemical being used when studying the effect. Paracelsus studied the effect of poisons on organisms. He astutely observed that many seemingly innocuous chemicals had adverse effects on organisms when given in very high doses. More importantly, he observed that known poisons had little effect on organisms if given in very small amounts. These thoughts were summed up in the phrase 'the dose maketh the poison'. Hence, when judging the *biological activity* of any chemical, a reference point is needed as to what concentration should be used in making a judgement. Furthermore, because each species of organism differs from each other, a particular chemical might show some *biological activity* against one organism at a specified concentration but have no effect on many others. It follows from these considerations that it is predictable that if one tests a particular chemical at a very high concentration in a very wide range of organisms, there will be a high probability that that chemical will be found to have *some* effect on *some* organism under *some* conditions. However, as one reduces the dose of each chemicals being assessed, the fraction of chemicals that produce an observable effect will reduce. Similarly, if one reduces the range of organisms being used to assess the biological activity, the proportion of chemicals showing an observable response will fall. Finally, as Paracelsus would have predicted, if one assessed the effect of low doses of some chemicals on a quiet specific effect on only one species, the chances of finding that any one chemical showing biological activity would be very low indeed. So, the chance of any chemical possessing 'biological activity' depends entirely on the way in which one is measuring biological activity. Without specifying the dose being used and the breadth of the organisms being challenged in an assessment, biological activity can mean very different things to different people.

It was this flexibility in the definition of the term central to the argument about the role of NPs that covered the most glaring deficiency of the model—the great majority of NPs have never been shown to possess any form of biological activity against even one organism. Was this because, as advocates of the model argued, the real targets of the majority of NPs had never been identified? For example, a scientist studying the importance of members of a particular class on NPs as insect defence chemicals would not be concerned if only one particular member of this family of NPs seemed to possess potent insecticidal activity against one insect species because that researcher could assume that other researchers would find roles for all the other members of this class of NPs, either defending the plant against another species of insect, against one of the many species of fungi, against bacteria, against nematodes or indeed against other species of plant. Once again the fragmentation of the study of NPs encouraged people to be highly selective with the data being gathered. There simply was no collective responsibility to assemble a coherent model to explain the NP diversity in all organisms.

Remarkably, the clear prediction that all NPs should show unambiguous evidence of some form of biological activity, at concentrations that the producing organisms could realistically achieve, was testable. Indeed, the prediction was being tested daily for the past three decades of the twentieth century by many large pharmaceutical companies around the world as they sought new drugs or by agrochemical companies searching for new insecticides, herbicides and fungicides. Although for commercial reasons, most of these data were not widely available (see Chapters 4 and 7), sufficient data were published to allow some judgement to be made about the probability of any one chemical possessing biological activity.[14] For example, it has been estimated that in excess of 100,000 different, related chemicals (based on the structures of some powerful insecticides—organophosphates) were synthesised by agrochemical companies seeking new insecticides but less than 100 of those chemicals were of commercial value. One agrichemical company (ICI) reported that the chance of finding a single useful product was 1 in 1800 in the crude tests (on whole organisms) being conducted in 1956 and fell to 1 in 15,000 by 1978 when more specific forms of activity were sought. The large-scale screening for biological activity by pharmaceutical and agrochemical companies provided overwhelming evidence that the probability of finding one strikingly potent chemical was very low indeed—often <0.01%. Because this evidence came from the screening of synthetic chemicals, some questioned its relevance to NPs (read Chapter 4 to explore this assertion more fully). However, the large-scale screening of collections of NPs, although conducted much less often, provided very similar but low hit rates.[14] For example, one screen of the NPs made by 10,000 different microorganisms found only one clinically effective agent. If every plant was making a natural insecticide, why had so few potent natural insecticides been found?[17] If every plant relies on making an NP with fungicidal properties for its defence, why have so few fungicidal NPs been found?[17] So, all large-scale screening trials, whether conducted using synthetic or naturally made chemicals, have shown that the probability of finding a chemical that can specifically target one biological process is extremely low. If we now consider the molecular processes that allow chemicals to bring about their biological effect, this low hit rate can be explained.

The interaction of molecules determines the interaction of organisms—the concept of biomolecular activity

All large processes are governed by the properties of the lowest level interactions occurring in that process. For example, the greenhouse effect (the trapping of solar energy by the earth's atmosphere) is a property of the earth's atmosphere but the characteristics of the atmosphere are the result of the properties of the individual molecules that make up the gases in the atmosphere. Likewise, the properties of those molecules are dependent on the properties of the elements that make the molecules. And so on until one reaches the most fundamental particle that makes up each atom. So, the properties evident at a global scale are highly dependent on properties at lower levels of organisation—the properties are inherited as one goes from the fundamental particle to the largest scale being analysed.

This same concept of inherited properties applies to biological systems. At each level of biological organisation—atom, molecule, organelle, cell, tissue, organ, organism, population and ecosystem—constraints will have been imposed on the higher levels of organisation by each of the properties of all levels below. Because evolution works by selection on options that are available, chemical and physical constraints will be significant constraints on those options. When considering the chemical interactions between organisms, the constraints imposed by the way in which chemicals interact with each other at a molecular level will have been a fundamental constraint on what is seen at the organism level. The interaction between two highly evolved organisms will be no different from the interaction between two very simple organisms in terms of what happens at the level of the interaction of chemical A with chemical B. The more highly evolved organism might process the information it gains by detecting chemical A interacting with chemical B differently from the simple organism, but the basic constraints imposed by the way in which A can interact with B remain. This is why it is important to build any theory about the evolution of NPs on what is known about the way in which molecules interact, because the constraints imposed by those interactions will have been inherited over the billions of years during which NPs have evolved. Whatever effects might be measured in higher organisms that evolved billions of years later, those effects will still be governed by the same ancient molecular rules. This is why the term 'biomolecular activity' was introduced to help understand how NPs evolved:[18]

The biomolecular activity of a substance is the ability of that substance, when present at a low concentration, to significantly influence the function of a specific protein.

The advantage of thinking in terms of the biomolecular activity of substances rather than their 'biological activity' is that it is more discriminating. For example, there are hundreds of substances that show 'biological activity' as insecticides but not all chemicals classed as insecticides kill the insects in the same way (there are many ways to kill an organism). One can be more specific by classifying insecticides on the basis of their mode of action. For example, a small subset of insecticides kills insects by interfering with the insect's acetylcholine esterase enzyme and so the vague term 'biological activity' (insecticide) can usefully be replaced by the categorisation based on a specific biomolecular activity (acetylcholine esterase inhibitor). It follows that one form of generalised 'biological activity' may encompass one or more forms of biomolecular activity.

What do we know about the basis of biomolecular activity?

The study of dyes might seem an unlikely way of beginning a process of probing the way in which chemicals cause their biological effects. However, pioneering studies of the way in which different dyes stained cells (see Figure 6.2 and Chapter 7) provided some important clues about the fundamental processes. Paul Ehrlich (this Paul Ehrlich was a German medical researcher, who lived between 1854 and 1915, and should not be confused with the other Paul R Ehrlich of the Chemical Co-evolution Model) and others showed that there was usually a great specificity of the interaction between any one dye and some parts of the cells which they stained. This specificity was associated with the chemical structure—for example, some red dyes might stain one structure in one type of cell, but other red dyes would not do so. This tells us that there is *specificity* in any interaction between chemicals. It was also noted that the degree of staining depended on the concentration of the dye used—pale colours were produced by low concentrations and denser one by higher doses. It was also noted that the strength of the association between a dye and the specific component of the cell to which the dye 'bound' varied; for example, some dyes were easily washed out of a stained cell once the dying solution was replaced with a solution free of dye but other dyes were harder to remove by washing. Such observations tell us that the interactions between some dyes and their targets are *reversible*. It also tells us that the *rate of reversibility* is not fixed but a variable. These principles became clearer when new types of biological activity were explored in the early part of the twentieth century. Ehrlich used the concepts of specificity to guide him in seeking selective toxins. By investigating the effects of hundreds of individual chemicals on the viability of bacteria and higher organisms, it became clear that many chemicals were not sufficiently toxic to the bacteria to give them any potential as agents to kill bacteria but a few chemicals did have potent antibacterial properties and an even smaller proportion of chemicals inhibited the growth of bacteria but were tolerated by higher organisms. So, the numerous studies carried out on the way in which chemicals interact with substances in cells, starting with dyes, then continuing with pharmaceutical drugs, insecticides, fungicides, herbicides and endogenous hormones have produced some basic features that can be said to characterise the way in which a small molecule interacts with a large molecule to produce a biological response.

Interactions are highly specific. Of the thousands of molecules that enter your nose as you walk around a garden, only one, would smell of peppermint. When your doctor gives you penicillin, it will not harm your cells and it will not even kill all the bacteria in your body but it will kill those responsible for your treatable infection. If you want to kill the broadleaved weeds in your lawn, your garden centre will sell you a product that will target those weeds but which will not kill the grasses that make up your lawn. These examples of specificity underpin the pharmaceutical, agrochemical and veterinary drug industries and to some extent the food and fragrance industries.

There is a known relationship between the dose of a chemical and the magnitude of the response generated. When studying the effect of a chemical on any biological system, there is a characteristic relationship between the amount of a substance administered and magnitude of the response induced. This relationship is called the dose–response curve (Figure 5.1).

Most biological effects are reversible in the short term. The majority of effects of chemicals on biological systems are to some extent reversible in short-term experiments. For

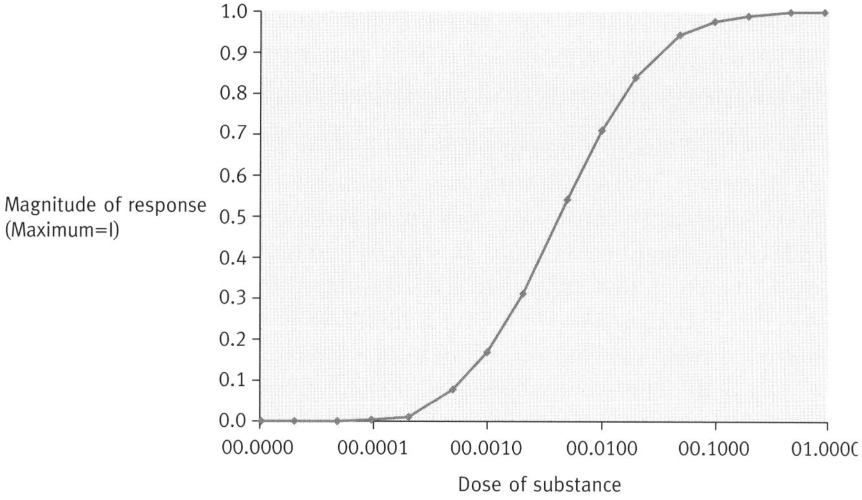

Figure 5.1. An ideal dose–response curve. Note the horizontal *x*-axis is plotted on a logarithmic scale while the *y*-axis is plotted on a linear scale. Consequently to double the response from 20% to 40% requires very much greater than a doubling of the dose applied. A consequence of this is that when gaining such data experimentally, it is much easier to gain statistically sound data where the curve is steepest (20–80% maximum response) than at the extreme points of the curve where large changes in the dose cause little change in the response. When doing an experiment on whole organisms to produce such a dose–response curve, this ideal relationship is sometimes not found due to the fact that the applied substance might have difficulty gaining access to the site of action, metabolism (the rate of which will have a different dose–response relationship) might be influencing the result and the ability of the organism to show a 100% response might be limited.

example, if one administers a sublethal dose of a chemical that inhibits the rate of multiplication of a bacteria growing on an agar plate, the effect will be lost if the chemical is washed from the plate—the bacteria will start to multiply again once they are no longer experiencing a significant dose. An example that most people will be familiar with is a local anaesthetic—its effect wears off after some time as the concentration of the chemical at the place where it acts decreases. The majority of pharmaceutical drugs and agrochemicals act reversibly.

These factors are each providing clues about the way in which small molecules (roughly speaking chemicals with a molecular weight of less than 1000) cause their biological effects if indeed they show any effect. One well-established chemical law, The Law of Mass Action, underlies these findings and, as we shall see a little later, this law guides us to the new model of NP diversity.

The Law of Mass Action, binding sites and receptors—understanding why specific, potent biological activity is a rare property for any one chemical to possess

Binding sites. Why is specific, potent biological activity a rare property? An important clue came from biochemists studying how organisms transformed chemicals using enzymes. Enzymes had been shown in the nineteenth century to be proteins. In the early decades of the twentieth century, the interaction of the enzyme with its *substrate* (the chemical on which the enzyme acted) suggested that every enzyme must have a unique structure that allows it to 'bind'[19] very tightly to its substrate. Part of the enzyme 3-D structure has been selected by evolution to complement the 3-D structure of the substrate. The better the complementation between the enzyme and substrate the tighter the binding (in other words, the enzyme will bind a substrate at even lower substrate concentrations). When biochemists studied the ability of enzymes to accept different, structurally related substrates, they usually found that most enzymes involved in the basic metabolism (see Chapter 9) were very *substrate specific*—the enzymes only accepted a very limited range of substrates. Furthermore, some chemicals very closely related to a substrate would bind to the enzyme quite strongly but the enzyme could not transform the chemical into a new structure and such a chemical, if added together with the normal substrate, would inhibit the enzyme action on its normal substrate. Such inhibitory *substrate analogues* sometimes found a use as *competitive inhibitors* to reduce the capacity of that enzyme experimentally. These interactions between the binding sites on enzymes and the small molecules that associate with those binding sites became the models for the way in which any small molecule associated with a protein in a specific manner.

In the 1960s, physiologists became interested in how hormones could bring about their effects in organisms, even though the concentrations of the hormones in the cell were often extremely low (micromolar or nanomolar). Because of the importance of hormones in human physiology, and because people with abnormal hormone physiology show serious health impairments (diabetes, thyroid deficiencies, etc.), close attention was paid to the way in which cells sensed hormones. Many chemicals had been made which were structurally similar to known hormones and their biomolecular activity (their ability to mimic the effect of the hormones) determined. It was clear from such studies that every type of hormone-sensitive cell had very great powers of discrimination; the cells would sense only a very limited number of hormones or their synthetic mimics. It was also known that the interaction between these cells and any added hormone was a reversible one—the effect of the added hormone could be reduced or abolished if the cell was exposed to a hormone-free solution. Because all these features were shared with the well-known enzyme–substrate interactions, it was postulated that hormone-sensitive cells must contain specific proteins (*hormone receptors*) which could associate with particular hormones. Because the hormone was unchanged by the association with the protein, the hormone was not described as a substrate but it was termed a *ligand*.

The discovery and full characterisation of several different hormone receptors led to a greater appreciation of the way in which cells could detect and respond to any chemical. This knowledge, taken together with the understanding of the way in which enzymes associated with substrates, made it clear that the association of small molecules with proteins followed a basic set of fundamental principles, principles that could usefully guide studies of the way in which drugs act on cells, the way in which herbicides influence plant growth and development, the way in which fungicides act on fungal cells and the way in which insecticides act in the cells of insects. In all these cases, the fundamental features of the effect of the biologically active chemical were reversibility, a very restricted range of structures showing high activity and an ability to bring about an effect even when the chemical was applied at very low doses. All these features were also characteristics of the way in which most NPs are known to act. Hence, it is reasonable to assume that the huge amount of information about ligand–protein interactions can inform our quest to understand the way in which NPs act on cells.

Quantitative relationships in ligand–protein interactions—the Law of Mass Action

The Norwegian chemists Cato Maximilian Guldberg and Peter Waage, between 1864 and 1879, proposed a means of quantifying the rate of a reversible chemical reaction between two substances. Given the reaction is reversible, at equilibrium, the rate at which chemical A reacts with chemical B to give AB is equalled by the rate at which AB decays to give A and B (Figure 5.2). These ideas where developed further by biochemists who recognised that the same principles applied to the reversible interaction of an enzyme and a substrate so that such interactions could be described mathematically in a similar way. The concepts were next applied to the interactions between drugs and the proteins and then to the interaction between hormones and their 'receptors'. The relationship first proposed by Guldberg and Waage, and developed by others, is the *Law of Mass Action*.

For the sake of illustrating our discussion, let's use the example of a hormone (H) interacting with its receptor (R) (hormones can be considered to be closely related to NPs, indeed some might be considered such as discussed in Chapter 9). Suppose we have several tubes each with the same volume of solution containing the same concentration of a hormone receptor protein and let us assume that we have a simple technique to measure whether each receptor is binding to a hormone at any one moment. Because the interaction of the hormone with the receptor is reversible, and follows the Laws of Mass Action, if one adds a different amount of H to each tube one can measure at what concentration of H there is no significant occupation of the receptors, at what concentration of H is half filling them and which concentration of H just fills all the receptors. Plotting these data on to a graph would provide the equivalent of a dose–response curve (Figure 5.3). This *in vitro* relationship could provide good clues as to the expected *in vivo* relationship in some cases. For example, if a cell containing the receptor could be

$$\text{Chemical} + \text{Protein} \underset{K_{\text{off}}}{\overset{K_{\text{on}}}{\rightleftharpoons}} \text{Chemical–Protein}$$

$$[\text{Chemical}] \times [\text{Protein}] \times K_{\text{on}} = [\text{Chemical–Protein}]\, K_{\text{off}}$$

$$\frac{[\text{Chemical}] \times [\text{Protein}]}{[\text{Chemical–Protein}]} = K_{\text{off}}/K_{\text{on}} = K_{\text{d}}$$

Figure 5.2. The Law of Mass Action is a simple mathematical relationship between the concentration of the protein which binds the substance, the concentration of the substance and the concentration of the protein currently binding the substance.

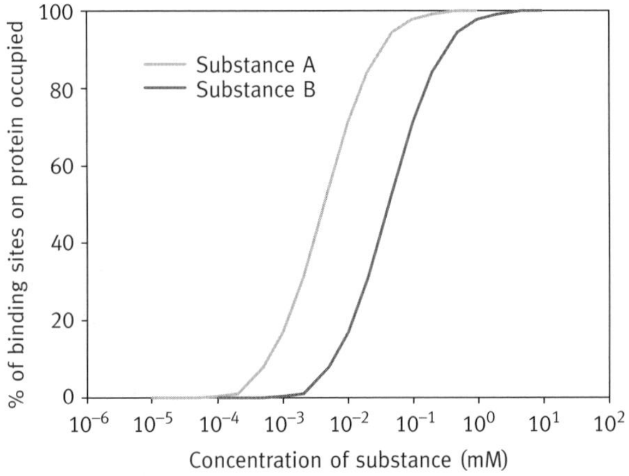

Figure 5.3. A binding curve is simply a dose–response curve determined at the biomolecular level. In contrast to the dose–response curve determined on a whole organism, it is easier to ensure that the substance has good access to the protein, metabolism can be eliminated and the full range of a response will be available. In this graph, the binding curves of two different substances capable of binding to the same protein are shown. Substance A occupies 50% of the binding sites at one-tenth the concentration needed to bring about the same degree of occupancy with Substance B.

shown to produce a hormone-specific response and if the hormone could freely enter and leave the cell and was not rapidly metabolised, one would expect the relationship between the concentration of H in the cell and the magnitude of the hormone response of the cell to follow the same curve measured using the purified receptor.

So, the Law of Mass Action is at the heart of all reversible interactions between small molecules (natural and synthetic) and proteins with which they interact to produce any type of biological response.

We can carry these thoughts a little further forward by thinking about what would happen if the *in vitro* experiment was used to compare efficacy of several different hormone analogues. Such studies usually find that each analogue has a unique relationship with the receptor. The shape of the curves usually does not differ but the position of the curve on the horizontal axis is characteristic of each analogue. The magnitude of the shift is directly related to the ability of the analogue to bind effectively to the receptor—analogues that fit the receptor poorly need a much higher concentration of free analogue in solution to half fill the receptor than analogues that better fit to the receptor (Figure 5.3). The final piece of useful information that these relationships tells us is that a binding curve usually goes from zero receptor occupied to full occupancy over about two orders of magnitude of the concentration of the active substance. For example, if one found that chemical X started to interact with protein Y when the [X] was 1 micromolar, one could predict that at 100 micromolar X, nearly all the proteins in the solution would at any one moment be interacting with an X molecule and any further increase in X would produce no extra effect (the response saturates).

The Law of Mass Action and the specificity of action of NPs

So the Law of Mass Action informs us about the basic rules that underlie biomolecular interactions. So, when studying some form of 'biological activity', this law is governing what is happening. Every cell contains thousands of proteins, each playing a specific role, so in theory each protein is vulnerable to interference by a few small molecules binding to some site on the protein's complicated 3-D structure. But the Law of Mass Action tells us that if the small molecule is present in solution at a very low concentration it will only be able to occupy sufficient sites on one specific protein if it has an exquisite fit with any vulnerable site. Thus, the theory behind the Law of Mass Action predicts that when tested at a low concentration, there is a very, very low probability of any one chemical structure possessing the right 3-D structure to effectively bind to any one type of protein. So, testing a low concentration of an NP for biomolecular activity against a specific protein target predictably will have a very low chance of finding a significant interaction. However, if tests of the NP are made using a cell-based, rather than a molecular-based, assay the chances of finding some form of biomolecular activity increases a little because each cell contains thousands of proteins. The assessment of 'biological activity' might be expected to produce a higher incidence of activity if the testing was conducted on a whole organism because an even larger number of potential protein targets would be aggregated in the assay. If the chemical is screened for activity in many species of mammals or insects for example, the chance of finding an effect increases only a little but because every species contains many highly conserved proteins; one is effectively retesting against very similar or identical proteins in each species. If one tests against a wider range of organisms, the chances of finding potent biomolecular activity increases a bit more but again some proteins are highly conserved across very diverse groups (Figure 5.4).

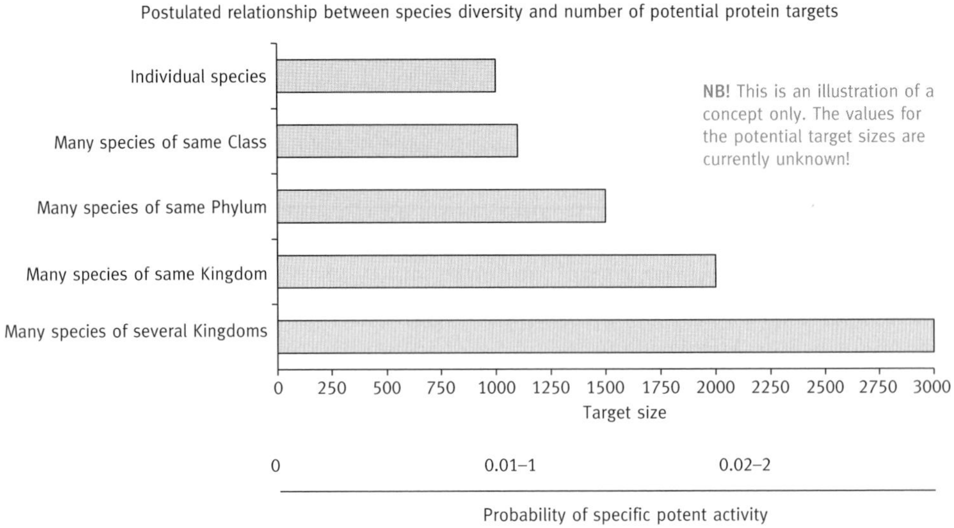

Figure 5.4. Theoretical relationship between number of target proteins and chance of finding biological activity at different organisational scales. The main point being made is that most organisms share a very large number of proteins that have been highly conserved, these proteins being very important to the short-term fitness of the organism. Consequently, if an applied chemical at a particular dose has a 1 in 1000 chance of reducing the fitness of one species then testing the same chemical, at the same dose, on 1000 different species of the same taxonomic class will only increase the probability of finding a chemical capable of reducing fitness very slightly. As one extends the range of species being tested, the probability of finding significant activity only creeps up. The figures given in this graph are invented for the purpose of argument.

The implications of the Law of Mass Action to the evolution of NPs—the Screening Hypothesis

Given there is a very low probability of any compound (whether made by a human or made by any other organism) possessing potent, specific biomolecular activity, how would this fact have influenced the evolution of NPs? One can build a simple model of the evolution of a metabolic pathway where each postulated step in the evolutionary process is assigned a probability and one can run the model to see how the possible evolutionary outcome would depend on the probability assigned to each process.[20] At its simplest one starts with one new NP (NP′) being made by a mutant in a population that happens to possess a new enzyme that can make NP′ from an existing substance X in the cell. One then adds another step in the process so that another mutation arises, which can make yet another novel NP″ from NP′ (Figure 5.5). One can assign a probability for each new NP having a type of biomolecular activity that is beneficial to the producer but one also has to consider whether another type of mutation might bring

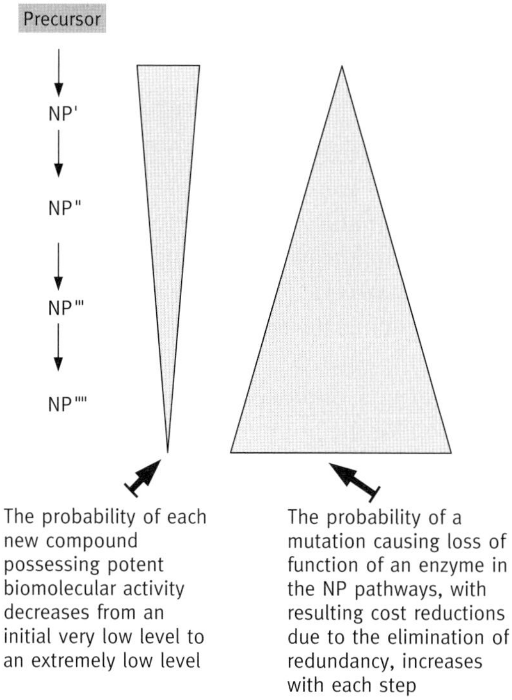

Figure 5.5. A diagrammatic representation of the theoretical chance of extending an NP pathway. The NP pathway starts with a mutational event allowing the organism to make NP′ from a precursor which is being made as part of the repertoire of chemicals needed by the cell for basic functioning (see Chapter 9). The chances of NP′ possessing useful biomolecular activity is very low and if NP′ has no useful biological activity subsequent mutational events have a greater chance of ridding the organism of NP′ than producing another new enzyme capable of producing NP″ (because hundreds or thousands of mutational events could destroy the activity of the enzyme making NP′ but only a handful of mutations might result in the production of an enzyme capable of making NP″ from NP′). The disparity between the chances of a mutation causing a fitness gain by 'cost savings' instead of a fitness gain due to the production of a new form of biomolecular activity grows as the NP chain lengthens.

about a benefit in a mutant by ridding the mutant of some unnecessary costs. In other words, the chances of a new pathway extension depends not only on the probability of any new substance made bringing significant benefits to the producer, but also on the chances of a mutant arising that saves costs by ridding itself of a redundant or useless enzyme (or indeed a whole pathway). Because it is more likely that a mutation will abolish an enzymic activity rather than produce a useful novel form of enzyme, the potential cost savings by pruning biochemical dead wood will be significant. The reason that the loss of function of a protein by mutation is much more likely than the production of a useful novel protein, are that most changes to a protein structure are likely to be detrimental because evolution has already optimised the protein's structure (see Chapter 9). Thus, there are several probabilities to be considered when thinking about the evolution of NP-producing pathways. At each stage of the pathway extension, the probability of the novel NP possessing useful biomolecular activity remains very low, yet the opportunities for cost savings increases at each stage. Furthermore, it could be argued that the probability of gaining useful biomolecular activity decreases with each step because any new NP made will be structurally related to an existing NP, an NP that the target species will already be co-evolving to resist or tolerate. Studies of the development of resistance to synthetic control agents have revealed that many resistance mechanisms cope with chemically related substances. So, how have NP pathways evolved?

The Screening Hypothesis,[13] first outlined by Jones and Firn in 1991, made the radical proposal that, to compensate for the poor odds demanded by the Law of Mass Action, organisms that gain fitness by making NPs must have evolved to generate and retain chemical diversity. The hypothesis was named to emphasise that organisms making NPs face the same challenges as humans trying to find new drugs or agrochemicals using screening trials. Humans designing a screening trial know that the more chemicals they test, the more chances there are of finding a useful chemical. Humans also know that keeping as much chemical diversity as possible to use in another screening trial, often seeking an entirely different form of biomolecular activity also makes sense.

So, how could NP-producing organisms increase their chances of making and retaining chemical diversity? The Screening Hypothesis proposed that by evolving certain metabolic traits, most importantly abandoning the dogma taught in every elementary biochemistry course or textbook, that enzymes are always substrate specific, one could indeed predict that there could be ways that organisms could maximise the production of chemical diversity and increase the chances of retaining even some 'redundant' NPs. The retention of some 'redundant' chemical diversity is necessary in order to seed the generation of new chemical diversity.

Enzyme specificity

When students learn about enzymes, it is nearly always stated, very dogmatically, that enzymes will act on only one chemical—they have very narrow substrate specificity. Indeed, when introducing the concept of enzyme action at an elementary level, the analogy of the lock and key is often used to illustrate the concept of specificity. In other words, students are told that for every product found in a cell, there will be one enzyme that has made that product and that enzyme will make no other product. This idea was extended to the idea of one gene–one enzyme–one reaction. There is indeed a considerable body of evidence to support the view that many enzymes do have narrow substrate specificity. However, exceptions were known to this 'rule'. But more importantly, the types of enzymes used as good examples of the 'lock and key' concept were drawn from a rather limited range of examples, mainly examples of enzymes that were involved with the basic metabolism found in a wide range of organisms (see Chapter 9 for a more detailed discussion).

Why do so many well-studied enzymes have narrow substrate specificity? Is such specificity inevitable? Firn and Jones[21] (Chapter 9) have argued that there is no *a priori* reason to assume that all enzymes must be substrate specific. They have suggested that high substrate specificity, rather than being inherent, will usually be the result of evolutionary selection. When a mutant organism produces a protein which possesses a novel enzyme activity, any benefit that new enzyme might bring to the organism must arise not from the properties of the enzyme but from the properties of the product that the enzyme produces. A novel protein can only be regarded as an enzyme if there is a substrate available for its action. A mutated enzyme that produces no product imposes the cost of its production on the producer but there can be no benefit; hence, the mutant will be lost from the population. Consequently, new enzymes that possess a very broad substrate specificity will have a greater chance of producing a new chemical, hence have some chance (albeit very small) of benefiting the producer. A new enzyme that can only act on one substrate has a much higher chance of being operationally inactive; hence, such mutants would, on average, be lost from the population. However, once a mutant has gained fitness by possessing a new useful chemical, evolution can act to optimise the synthetic processes leading to that chemical. A subsequent mutation leading to a modified enzyme with a narrower substrate specificity will result in fewer unwanted products and more of the desirable product; hence, the new mutation might under some circumstances be favoured by selection. Thus, enzyme specificity is very much an evolved characteristic and as explained in Chapter 9, the selection forces to bring about this narrowing of substrate specificity may be huge for one type of metabolism (*integrated basic metabolism*) but very small in another type of metabolism (e.g., NP-producing metabolism).

So why would a broad substrate specificity enhance the generation and retention of chemical diversity?

Consider the pathway shown in Figure 5.6. The upper panel shows the traditional view of the one enzyme/one product pathway. However, suppose that we relax substrate specificity such that each of the three enzymes can act on any substrate, as shown in the lower panel. The result is that the order in which the changes are made to the starting product becomes unimportant and many different structures can be created by the same three enzymes. Broad substrate specificity not only increases the generation of chemical diversity but it does so at low cost. Because making an enzyme is expensive (a large number of amino acids are needed for every enzyme molecule and the machinery used to make the enzyme consumes energy), this is a significant saving and would predictably help compensate on the balance sheet for the 'waste' costs of making some chemical diversity that serves no useful purpose.

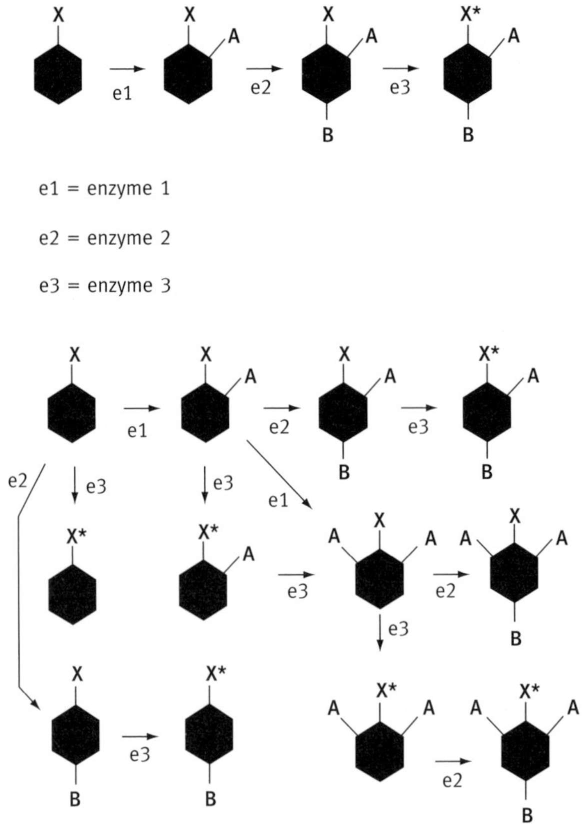

Figure 5.6. The production of NPs using 'matrix pathways' was predicted by Jones and Firn[13] because of the opportunity to produce and retain chemical diversity efficiently. In this diagrammatic scheme, three enzymes (e1, e2 and e3) have access to one substrate. The upper panel shows that if each of the enzymes has a strict substrate specificity, a linear pathway producing three new chemicals would be expected. However, if the three enzymes have a broad substrate specificity then the order of conversion can vary and a matrix pathway will result. Now three enzymes will produce 11 novel substances. Furthermore, such matrix pathways are more robust to the loss of any one enzyme activity (see Figure 5.4).

Broad substrate tolerance would also help ensure that some chemical diversity would be retained. In Figure 5.6, if any of the product that requires the action of all three enzymes for its synthesis endows the producer with enhanced fitness, then the other products with no role would be made initially and would only be lost by subsequent selection. Clearly, there is a balance between the advantages of generating chemical diversity, the advantages of retaining chemical diversity to beget future chemical diversity and the disadvantage of retaining costly redundancy.

Evidence to support this proposition that enzymes involved in NP metabolism would possess broad substrate tolerance

When, in 1991, the prediction was made that enzymes involved in NP biosynthesis would have broad substrate specificity, there was not a large amount of evidence available to support the proposition. There were possibly several reasons for this paucity of evidence.

1. Researchers tend to find what they seek; hence, if the prevailing view among biochemists was that enzymes were always substrate specific, seeking evidence to challenge such dogma would not be a high priority.
2. It is often difficult to isolate and purify many enzymes involved in NP biosynthesis (many organs or parts of organs rich in NPs often contain material which, released from the cell compartments when the tissues are ground up, inhibit enzyme activity). Only later when it became possible to isolate genes coding for NP-producing enzymes, did it became possible to use molecular biological techniques to add the appropriate gene to an organism that was easier to study. By choosing an organism with a simpler NP composition it became possible to identify more easily the minor products that the enzyme under study might be making.
3. Enzymes that make NPs necessarily act on complex molecules. Such chemicals can be very hard for chemists to synthesise; hence, the range of substrates available to the biochemist to assess substrate specificity was often limited.
4. The techniques available to isolate and characterise the minor products made by an enzyme were insufficiently sensitive or specific.

In the decade after the Screening Hypothesis predicted that some enzymes making NPs might have a broad, not narrow, substrate specificity, extensive evidence was indeed found which was consistent with the prediction.[22] The prediction was never made the focus of a major study; hence, the evidence had to be gleaned from results being published by those researching other aspects of NP metabolism. The problem with seeking evidence indirectly in this way is that there is a tendency to find the evidence being sought. However, even allowing for this possible bias, there is now substantial evidence that some enzymes involved in NP synthesis do indeed have a broad substrate tolerance.[23]

Mutants with changed NP composition

There are many garden plants that are closely related but have very different, and very characteristic, smells or tastes. The scented leaf geraniums (*Pelargoniums*), for example, can smell of apple, peppermint, cedar, rose and lemon. Most garden centres sell many type of mint (apple, ginger, peppermint, etc.) which also have very different characteristic smells and it was a study of mutants of mint (*Mentha*) that gives one of the nicest, and most complete, examples of the way in which several enzymes in a sequence leading to an NP pathway can readily accept new substrates. Researchers were seeking cultivars of spearmint that were resistant to a fungal pathogen that was reducing the commercial production of this plant in the United States. A mutant was found that showed an increased resistance to fungal attack but, unexpectedly, this mutant smelled like peppermint. The very characteristic smell and taste of spearmint is the result of the mixture of monoterpene (see Chapter 3) made by that plant. So, a comparison was made of the monoterpene composition of the plants that had smelled like spearmint and plants that had smelled like peppermint.[24] This analysis showed that the two types of plant made different, but related, mixtures of monoterpenes. Yet the mutant was known to be the result of a change to one gene; so, how had the change to the activity of one enzyme brought about several changes in chemistry? It was found that the mutated gene was an enzyme that added a hydroxyl group to the 3-position of the cyclohexene ring of limonene while the wild-type hydroxylated the 6-position (Figure 5.7). All the other new products made by the 'peppermint' were the result of the fact that all the downstream tailoring enzymes in the peppermint accepted the new 3-hydroxy substrates to give an array of new products. Clearly, the downstream enzymes were able to accept 3-hydroxy substrates or 6-hydroxy substrates. The appearance of an unexpected novel product in the peppermint suggests that a further elaboration of one of the newly created products by some unidentified NP enzyme, from another pathway, generated further chemical diversity. This is a fine example of how a single gene mutation can generate several new products and how new diversity can be propagated in unexpected ways.

Matrix pathways

If some enzymes involved in NP biosynthesis can accept more than one chemical as a substrate, the traditional view that metabolic pathways will be linear sequences becomes questionable. The Screening Hypothesis predicted that pathways would be found where the same enzyme might participate in two different linear sequences. The hypothesis also predicted that there might be more than one route to a given product depending on the sequence of enzymic steps—as illustrated conceptually in Figure 5.6. Both of these predictions have been verified in biosynthetic routes to a number of NPs. One such example comes from studies of the pathways leading to flower colours (see Chapter 9 for a discussion as to whether plant pigments are NPs), in particular the *anthocyanoside* flower colours. For example, in petunia flowers, three enzymes (F3H, F3′5′H and F3′H) can produce five different products (eriodictyol, pentahydroxyflavone, dihydromyricetin, dihydroquercetin and dihydrokaempferol) from naringenin. Studies of the flavonoid 3-*O*-glucosyltransferase (3-GT) in *Perilla* also provide evidence for its role in a metabolic grid. Another example of a metabolic grid is found in the synthesis of lignin, the complex material found in plant cell walls (see Chapter 9 for a discussion as to whether some plant cell wall constituents should be called NPs). A scheme for monolignol biosynthesis has been proposed, where the three enzymes CAOMT (caffeic acid *O*-methyltransferase), CCoAOMT (caffeoyl-CoA *O*-methyltransferase) and hydroxycinnamate CoA ligase have sufficient substrate tolerance that they each act on more than one substrate to create a matrix of transformations. The compounds that

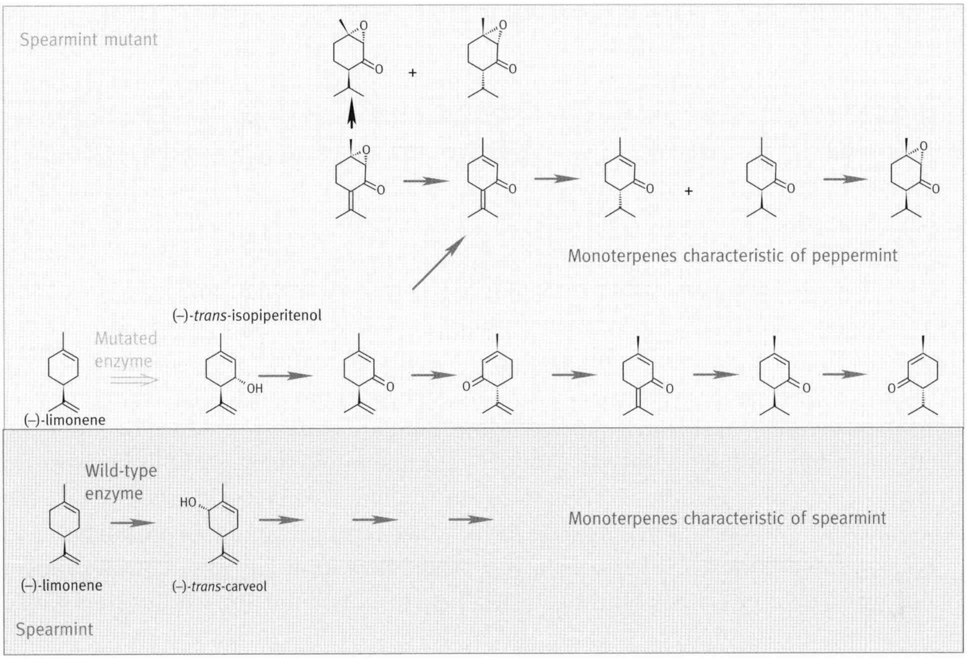

Figure 5.7. There are many examples now known of the synthesis of NPs via matrix pathways (see also Figure 9.3). However, a nice example of the benefit of such flexibility was revealed when a mutant of spearmint that had smelled more like peppermint was studied.[24] A comparison of the terpenes in both plants revealed that the single gene mutation had not resulted in a single chemical change but multiple changes. In the mutant plant, a hydroxyl group was added to the 3-position of the cyclohexene ring of limonene while the wild-type hydroxylated the 6-position. Some of the other wild-type tailoring enzymes in the mutant did not discriminate fully between the 3- and 6-hydroxylated products so a new family of NPs were produced which gave the mutant plant an odour of peppermint.

give the brassica crop products (cabbage, brussel sprouts, broccoli, cauliflower, salad rocket, etc.) their characteristic smells and flavours come from a group of chemicals called the glucosinolates (see Chapter 3). In excess of 100 glucosinolates are known and this diversity of aliphatic glucosinolates is thought to result from a grid of conversions using a limited number of enzymes involved.

Chemical reactions or rearrangements

One of the key differences between chemical methods of making compounds and enzymic syntheses is that the use of chemical reagents tends to create a number of products instead of the usual single product produced by enzymes (see Chapter 4). However, evolution has come up with a way that enzymes can exploit the fact that certain unstable molecules will rearrange to produce multiple products. There are only a few such enzymes currently known to catalyse the synthesis of such unstable molecules but they do generate a wonderfully impressive range of products. A study of the members of the Tpsd gene subfamily in Grand Fir (*Abies grandis*) found five monoterpene synthases (ag6, ag8, ag9, ag10 and ag11) that were capable of producing multiple products.[25] The most striking evidence for the ability of enzymes to produce multiple products comes from a study of two sesquiterpene synthases, also in Grand Fir. One enzyme (δ-selinine synthase) produced 34 different compounds from a single substrate and another (γ-humulene synthase) produced 52 products from its precursor. A further interesting example of the generation of multiple monoterpene products comes from a study of monoterpenes in Common Sage (*Salvia officinalis*), where the search for a (+)-bornyl diphosphate synthase and a (+)-pinene synthase led to the suggestion that both these activities reside in a single enzyme. Similar flexibility is shown by limonene synthase which has been shown to produce multiple products in isotopically sensitive branching experiments and cDNA cloning. In tomato (*Lycopersicon esculentum*), the sesquiterpene synthase germacrene C synthase also produces multiple products.

Clearly, the different synthetic capacities available within any one natural product pathway will determine the ability of an organism to exploit the opportunity to produce multiple products and the fact that the examples given come from the terpenoid pathway suggest that this strategy could be less universal than the selection of enzymes with broad substrate tolerance. The production of multiple products by an enzyme makes the naming of the enzyme somewhat difficult. Naming the enzyme after the major product produced seems logical but such a convention is arbitrary and even misleading. The product that enhances the fitness of the organism need not be the major product and evolution is most likely selecting the overall properties of the enzyme (and indeed the overall pathway), not the one product. A convention to denote the ability of an enzyme to produce multiple products would seem to be desirable.

How do the patterns seen in the NP pathways, described in Chapter 3, fit with this model?

Very well indeed, at the time when the Screening Hypothesis was proposed (in 1991), the patterns outlined in Chapter 3 (and summarised conceptually in Figure 3.2) were not clear. Since that time it has been found that, not only do individual NP pathways show many of the metabolic traits predicted by the Screening Hypothesis, but also taken as a group, NP pathways share a common strategy—to enhance the production and retention of chemical diversity at low cost. The predicted features enhance the chances of a single gene mutation in an NP pathway will result in more than one new NP being made.

Criticism of the Screening Hypothesis

Any scientific hypothesis is simply an attempt to provide a theoretical framework that allows people to understand the processes that interest them and also enables them to make predictions which can be tested experimentally. However, just because a hypothesis can provide a more complete explanation of observations than previous hypotheses, or makes predictions that are subsequently experimentally verified, does not mean that the hypothesis is valid. Those proposing a hypothesis, and those extolling its virtues, will inevitably focus the attention on the strengths of the hypothesis. Yet it is the weaknesses of a hypothesis that should really be the centre of attention. Critics, who point out the weaknesses of a hypothesis, do a service where they can contribute to the rejection, or the strengthening, of the model. Many critics of the Screening Hypothesis have unknowingly contributed to its development. These critics gave their views at conferences, as anonymous referees of papers or in discussions. Berenbaum and Zangerl[26] were the most focused and forthright critics and their views are discussed below because their criticisms covered all of the themes which preoccupied other critics.

Is biological activity a rare property for a molecule to possess?

Given that the most widely accepted paradigm at the end of the twentieth century to explain the evolution of NPs was based on the assumption (usually unstated) that all NPs were biologically active, Jones and Firn's questioning of this assumption was inevitably provocative. Berenbaum and Zangerl argued that the vague definition of the term *biological activity* was the fundamental flaw underlying the Screening Hypothesis.[27] By focusing attention on this issue, the critics served their purpose because Firn and Jones were forced to build a more substantial argument based on the concept of biomolecular activity.[22] The huge amounts of data gathered in the 1990s using cell-free screening (see Chapter 7) showed that for any one type of biomolecular activity, many thousands of randomly selected molecules would have to be screened in order to find one which had high, selective, activity. Furthermore, the reasons for the low probability of possessing potent, specific biomolecular activity could be understood in terms of the increasing knowledge of the way in which small molecules associate with proteins. Because it is impractical to assess the biological or biomolecular activity of every NP, sceptics may argue that this fundamental aspect of the Screening Hypothesis remains speculative. However, the argument can be turned round. Is there any evidence that supports the view that all NPs are biologically active?

Maybe the effects of individual chemicals is enhanced in the presence of other chemicals—the magic mixture argument?

NPs never occur in nature as pure compounds. The cells that make NPs are chemically complex. Even if an NP is released as a volatile or into the solution surrounding the maker of the NP, there will be other chemicals present. Berenbaum and Zangerl made another substantial criticism when they suggested that chemicals may be more effective in combination than when administered as a single substance. Given that most screening trials (but very importantly not all[28]) have been conducted using solutions of one chemical only, if mixtures of chemicals were more effective than single chemicals, this would make data from screening trials a very insecure basis for building an evolutionary argument about NPs. The challenge that such thinking presents to the Screening Hypothesis was clearly outlined by Berenbaum and Zangerl.[26] They argued that maybe individual NPs might show little biological activity when given to an organism in pure form but would show much greater biological activity when given in combination with other NPs. What experimental or theoretical evidence is there which addresses this key issue?

Using the concept of biomolecular activity, the argument can be progressed. If two chemicals show no biological activity alone but do when given in combination there must still be two individual biomolecular activities underlying the effect of each of the chemicals when they are applied in combination. The probability of each compound possessing potent biomolecular activity is not changed by the fact that another chemical is acting elsewhere in the organism. At the biomolecular level, there will be an independence of action except in the predictably very rare case when the two chemicals act on the same protein. Before exploring this further, one needs to consider how two chemicals could have potent biomolecular activity when combined but show little or no activity when assayed alone.

- One simple explanation would be that the effects of the two chemicals are simply additive. Suppose that two very similar drugs have identical modes of action. A patient might find that one tablet of A or one tablet of B was ineffective but one each of A and B had an effect, but so would two tablets of A, two of B. In other words, the combined effect is simply a consequence of the effective dose supplied. Both A and B will behave just like any chemical tested in isolation and the Laws of Mass Action will apply.
- An alternative scenario would be that one chemical X, with high inherent biomolecular activity, is rapidly degraded by the target organism. Owing to this degradation of X, an effective dose cannot be achieved. Suppose, however, that chemical Y inhibits the enzyme that degrades X. If Y is supplied with X, an effective dose of X is achieved. Thus, X or Y might both possess potent biomolecular activity but the activity of each only reveals itself in the presence of the other. Once again X and Y will behave just like any chemical tested in isolation and the Laws of Mass Action will apply to each chemical interacting with the protein with which it interacts.

There are other such scenarios that can be proposed. Evidence exists for both additive and synergistic actions.[29] There is nothing magical about mixtures. The effects of mixtures can be explained in terms of the actions of the individual chemicals, all of which obey the usual physicochemical laws.[30]

So how do these arguments apply to the evolution of NPs? Suppose that one chemical with rather weak biomolecular activity exists in an individual, but that activity will only become evident if another type of biomolecular activity is evolved. The probability of the first chemical possessing any form of biomolecular activity would be low. However, the chance of the combined activity becoming evident is the same as the chance of the second chemical possessing its form of biomolecular activity because one needs that second, very specific form of activity for the activity of the first chemical to be revealed. In other words, the magic mixture argument is an illusion. There is no escape from the fundamental Laws of Mass Action.

There are also reasons beyond the theoretical ones summarised earlier for questioning the idea that it is the mixture of NPs that are more important than any individual roles. Some screening trials have actually been conducted where mixtures of NPs rather than single compounds have been used. The testing of crude extracts of plant or microbial material has often been conducted and such trials usually show that there is a low probability of complex mixtures obtained by extraction showing potent, specific biological activity. Although there is evidence that the probability of finding activity is indeed greater in mixtures, rarely does such activity greatly exceed the predictable additive effects of testing multiple compounds.

Furthermore, it is often overlooked that thousands of screens of complex mixtures have been carried out inadvertently. Many tens of thousands of chemicals have been screened on whole NP containing plants for herbicidal, fungicidal or insecticidal activity. Each plant species will contain its own full complement of NPs. If adding a new chemical to a rich mixture of other chemicals really would increase the chances of finding biological activity, one would expect that such screens for pesticide activity would have a higher probability of success compared to screens using organisms with no NPs. Yet there is no evidence that such screens show higher probabilities of a particular form of biological activity than screens seeking activity in organisms that are devoid of NPs. Furthermore, it would be predicted that some very specific biomolecular activity would only be revealed when certain chemicals were tested on species of plants rich in a particular NP. The author knows of no evidence to support this prediction.

Furanocoumarins—good models or a group of chemicals with unusual properties?

The furanocoumarins are a group of NPs that are found in the wild parsnip, *Pastinaca sativa*, and in several other families of plants. In the nineteenth century, it was shown that the exposure of skin to this plant could cause severe skin irritation and blistering. Several scientists, in the early decades of the twentieth century, described similar symptoms but a key finding, by Kuske, was that blistering would only develop if the patient was exposed to light after rubbing the wild parsnip leaf on their skin. The chemicals in the plant that caused this photosensitised effect were identified as belonging to a group that were termed as the furanocoumarins (Figure 5.8). Because exposure to furanocoumarin-containing plants poses problems to humans and because some of the furanocoumarin family were thoroughly explored as potential drugs, an extensive literature developed. Interesting questions arose concerning the role of such NPs in the plants that made them and the way in which organisms that interacted with the furanocoumarin-containing plants coped with these clearly, powerfully, biologically active chemicals. The co-evolution of the furanocoumarin-containing plants and the insects

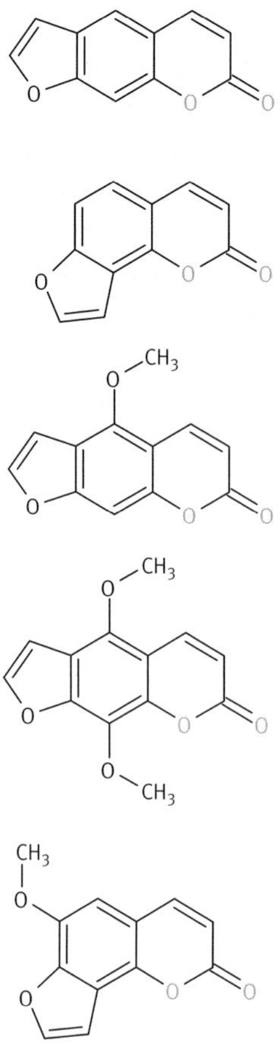

Psoralen

Angelicin

Bergaptol

Isopimpinellin

Sphondin

Figure 5.8. Some members of the furanocoumarins family of NPs showing the key feature (red) shared by each structure that allows each to react chemically with some nucleic acid and proteins. Possessing biological activity due to an inherent chemical reactivity is very rare among NPs for reasons explained in the text.

that could live on the plants was thoroughly studied by Berenbaum and her group.[31] Their data and experience were drawn upon when Berenbaum and Zangerl marshalled their arguments against the Screening Hypothesis. They presented evidence from the furanocoumarins that they argued showed that

- Many members of the furanocoumarin family are biologically active; hence, the probability of a molecule being biologically active was not low.
- Some members of the furanocoumarin family show more than one type of biological activity.

Arguing that the furanocoumarins were a good model for all NPs, Berenbaum and Zangerl proposed that their data presented overwhelming evidence against the

Screening Hypothesis. However, they overlooked the fact that the furanocoumarins were not representative of all NPs. Indeed, the furanocoumarins were very unusual among NPs wherein these chemicals were chemically reactive and they bring about their biological actions, not by reversibly binding to proteins, but by chemically reacting with not only proteins but also DNA. Because these chemicals depend on chemical reactivity, any member of the family of furanocoumarins that contains the appropriate reactive groups will possess some ability to react with biological materials hence will show biological activity. Thus, the probability that any member of the furanocoumarins family being biologically active is high because the group is partly defined by the very grouping that is important in endowing the chemical with chemical reactivity. The fact that furanocoumarins possess a broad range of biological activities is predictable because chemical reactivity is less selective than an activity dependent on reversible interactions with specific proteins. Thus, the furanocoumarins are good models for the way in which NPs with high chemical reactivity act but they are unrepresentative of the great majority of NPs which are not highly chemically reactive but act by reversible interactions with specific proteins.

Why are chemically reactive NPs not more common, given the evidence that evolution could have built families of related compounds, each compound having a high probability of possessing a broad range of biological activities? Interestingly, humans have also found it hard to exploit the apparently attractive properties of what were called 'active site irreversible inhibitors'.[32] It is possible to speculate that there might be high costs associated with the production of a reactive chemical, especially if that chemical has a broad spectrum of biological activity. The organism that must be exposed to the highest concentration of any chemical is the organism that makes the chemical; consequently, making a chemically reactive chemical is a high-risk activity. Any mutant that evolves the ability to make a chemically reactive chemical presumably has a higher chance of being less fit rather than being fitter. Furthermore, even if the producer of such a chemical does not suffer a loss of fitness, the lack of selectivity characteristic of chemicals, such as the furanocoumarins, means that any organism that interacts with the producer may suffer—beneficial insects or fungi would suffer along with deleterious insects or fungi. It would not be surprising if organisms have not often managed to evolve such that they gain the advantages of using chemically reactive NPs, yet bear none of the costs.

What does this chapter tell us about the way science works?

The century old debate about the roles and evolution of NPs illustrates nicely that the fragmentation of a subject can very significantly hinder scientific progress. The failure to construct a satisfactory universal model to explain the evolution of NP diversity was not due to a lack of effort or ideas or a lack of equipment or methodologies.[33] One clear problem was that most of the biochemists working on NP metabolism had been indoctrinated with dogmas (e.g., 'one gene–one enzyme–one product' or 'enzymes are highly substrate specific') that were simply inappropriate for enzymes making NPs because these dogmas were based on a type of metabolism that had evolved for a quite different purpose (see Chapter 9). The fact that the biochemistry underlying the most economically, socially and historically important group of chemicals had been relegated to this lowly position should warn society that scientists can be very blinkered.

What about the Screening Hypothesis? Exactly where the hypothesis stands in Schopenhauer's three stages cannot be fairly judged by one of the authors of the Hypothesis. However, even if the reader remains sceptical about its validity, the implications of the model to several other areas of biology, in the chapters that follow, provide interesting insights.

All truth passes through three stages.
First, it is ridiculed.
Second, it is violently opposed.
Third, it is regarded as self evident.
—*Arthur Schopenhauer*

6

NPs, Chemicals and the Environment

Sherlock Holmes ponders "the curious incident of the dog in the night time." Watson, surprised, responds, "But the dog did nothing in the night time." "That was the curious incident," Holmes replies.

—*Arthur Conan Doyle, Silver Blaze*

Summary

The world has never been a chemically clean place. The organisms that make NPs have been producing tens of thousands of different chemicals, totalling billions of tonnes every year, for billions of years. Hence, organisms have evolved to survive, indeed thrive, in the presence of these chemicals. The mechanisms evolved by organisms to cope with their chemical load are likely to be the same mechanisms used by organisms, including humans, to cope with new chemicals introduced into use by humans. Consequently, the subjects of toxicology and ecotoxicology can usefully be informed by the knowledge of NPs.

The fact that the world is not 'contaminated' by NPs to any significant extent, despite billions of years of this massive NP production, informs us that mechanisms must exist to recycle much of the carbon in NPs back into the global carbon cycle. The main routes of this NP carbon cycle must occur in microbes (which for billions of years were the only NP producers on the planet); hence, the ability of microbes to metabolise most of the substances made by humans is predictably drawing on this ancient metabolic ability. If one can understand the selection forces that could have resulted in this vital, useful and versatile metabolic capacity, it is easier to appreciate the capacity of individual organisms and ecosystems to degrade both natural and synthetic chemicals. Furthermore, by understanding the microbial capacity to degrade NPs, one can plan the best way of exploiting this capacity to target potentially harmful chemical pollutants.

The reader who is well informed about toxicology or ecotoxicology may wish to jump straight to the discussion of NPs in relation to these subjects.

The rise of public scepticism about 'chemicals'

In most developed countries, during the past four decades of the twentieth century, many citizens were becoming increasingly concerned about 'chemicals' in their food, in water and in 'the environment'. To the media, hence to the majority of the public, 'chemicals' referred to synthetic chemicals and especially to pesticides. The public concerns about the safety of the new synthetic chemicals being made and distributed around the world were not new but the concerns resurfaced in a new form and were more widely held. What have these public concerns about synthetic chemicals got to do with NPs? To most people, there is no obvious link between NPs and concerns about chemical pollution. Indeed, even to many scientists studying pollution NPs do not seem to be relevant to the subject—as can be seen by the lack of any reference to NPs in the popular books on ecotoxicology or environmental toxicology for example. However, as explained later in this chapter, only by understanding the way in which organisms have evolved in the presence of NPs over millions of years can one understand the impact that synthetic chemicals have on the natural world.

The changing attitudes to pesticides and pharmaceutical drugs

The acceptance of the government's role in protecting the public from hazards

Individuals had been aware of chemical pollution in all industrial countries since the nineteenth century, especially in areas near gas works, smelting plants, chemical factories, oil refineries or mines. In general, the poor and politically disenfranchised suffered most. The rich used their wealth to move to cleaner areas, in the same way they had for centuries fled areas contaminated by human and animal waste. In Europe, however, the replacement of numerous small local manufacturers throughout the country by a few huge factories in a small number of rapidly expanding cities increased the chances that significant numbers in the population would be exposed to harmful concentrations of toxic chemicals. Improving scientific knowledge, changing social attitudes and a broadening political franchise eventually allowed these growing problems to be recognised and addressed. The statutory control of the exposure of individuals to toxic chemicals in the workplace (e.g., the notorious effects of white and yellow phosphorous used in the manufacture of matches) began in the late nineteenth century and subsequently planning laws and emissions controls gave protection to whole communities that had previously been exposed to pollution.

These nineteenth century and early twentieth century great successes in improving the health of citizens, by the elimination or control of hazards, had a profound effect on the thinking of the public about risk. Good government, at national and local level, could increase the chances of individuals living a long and healthy life. It was the action of local or national governments, not individual behaviour, that had been most effective in improving the well-being of the average citizen, especially the poorer citizen. There was a growing consensus that it was the duty of government to protect citizens from hazards that the citizen could not control by their own individual behaviour. So, once

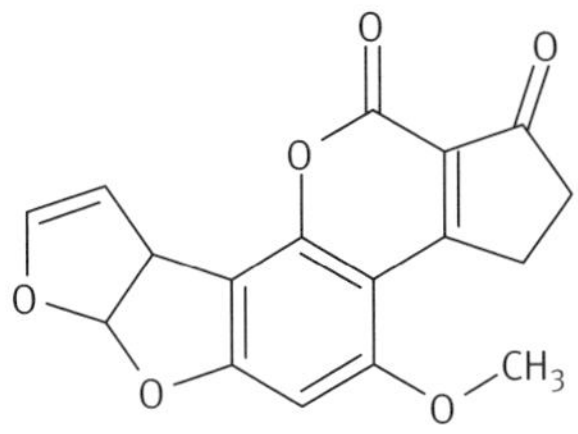

Aflatoxin B_1

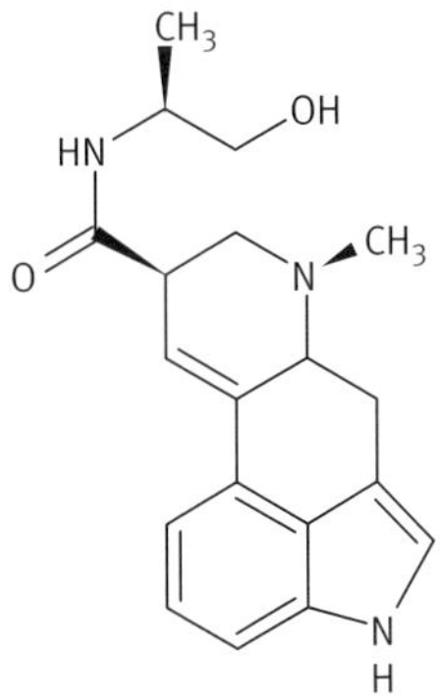

Ergometrine

Figure 6.1. Representatives of the Aflatoxin and Ergot alkaloid families.

safe drinking water had been made available to most citizens, attention was turned to food safety. In the nineteenth century, concerns about 'chemicals' in food were related to adulteration (reducing food quality by substituting low food value material) rather than the adverse effects that the minor toxic contamination might have on the consumers. There were some well-known examples of the chemical contamination of food, but it was not synthetic chemicals that were the problem but the contamination of grain with microbial NPs. The alkaloid ergot (Figure 6.1), produced by the ascomycete fungus *Claviceps purpurea*, induces hallucinations similar to those produced by the psychotic drug LSD (lysergic acid), it can induce abortions and can constrict blood capillaries to the extent that lethal gangrene may result from a lack of blood circulation. Rye, a major grain in northern Europe, was the grain that was most at-risk of ergot contamination but government-led control measures (improved analytical methods, good crop husbandry and improved processing) eliminated the problem. However, the contamination of food with toxic NPs remains a problem. Aflatoxin, produced by the mould *Aspergillus flavus*, can contaminate some nuts (especially peanut and Brazil nuts) and is converted by humans to a very powerful mutagenic chemical. Government-led food quality standards now stop the sale of aflatoxin-contaminated foods in many countries and even some bird foods are now advertised as being free of the toxin.

The rising concern about toxic chemicals in foods

It was soon after the widespread introduction of chemicals into agricultural use in the early part of the twentieth century (to kill insect pests, to control fungal diseases, to kill weeds and to rid animals of parasites) that public concern about the use of these highly poisonous chemicals in food production began to grow. These concerns now seem legitimate because the first generation of widely used agricultural chemicals included the salts of some heavy metals (arsenic, lead and mercury) that were very toxic to many organisms. There were also some toxic NPs (e.g., nicotine) used to control insects. The safe use of such non-selective toxic chemicals relied on the targeted application of the toxin directly to the insect or fungus in such a way that the harvested product was uncontaminated at the time of harvest. Clearly, this was relatively easy in the case of grain crops (where the grain developed well away from the leaves that might have needed chemical protection earlier in the season). However, this was very hard to achieve when the grower was trying to protect a product that the consumer would consume (e.g., apple, lettuce leaf, tomato, etc.), which was also the object of attack by the insect or fungus. Thus, the protection of the consumer relied heavily on the proper use of the poison by the user—always a recipe for disaster given the uncertainty of human behaviour. How could the public be sure that growers were not allowing poisonous residues to remain on the foods going to market? Recognising the potential for serious problems, in the early twentieth century, governments in many countries enacted laws in response to rising public concern. However, the seeds of doubt about the safety of agrochemicals had been sown in the minds of the consumers—the public realised that laws which rely on the unpoliced compliance of standards offer only limited protection. Public doubts were not helped by many lurid tales in popular fiction where murderers were often caught using rat poison, weed killer or some other noxious chemical, purchased legitimately, to send their victims to the grave.

The rising acceptance of the use of pharmaceutical chemicals

At around this time, in the first decade of the twentieth century, the German microbiologist Paul Ehrlich (1854–1915) showed that some synthetic chemicals had a highly selective toxic action on some organisms. Ehrlich took his inspiration from the accumulating evidence that some of the increasingly available synthetic dyes (see Chapters 2 and 8) stained only some natural substances. Some dyes only stained cotton but not wool for example. Microscopists had long used dyes to aid the visualisation of cells and organelles, using the synthetic dyes to identify cell types because the walls of some types of plant cell would associate with certain coloured dyes and not others (Figure 6.2 and read Chapter 9 for a possible explanation of why this should be so). Ehrlich was particularly impressed by the fact that synthetic chemical dyes stained some microbes and not others[1] and by the fact that certain parasites took up dyes easily. Ehrlich reasoned that it would be expected that organisms, tissues or organelles that accumulated dyes would in effect be exposed to those chemicals at higher concentrations than those structures that did not take up the dyes. Given that he knew that some dyes were toxic at high doses, he reasoned that selective toxicity might be possible. By screening[2] a large number of dyes, Ehrlich showed that it was possible to find a chemical that killed a parasite without killing the host.[3] The important concept of 'selective toxicity' was verified and the large-scale screening of large collections of chemicals, in the search for the rare chemical that might kill one type of cell and not another, was adopted by many companies, companies that eventually became part of the huge pharmaceutical industry

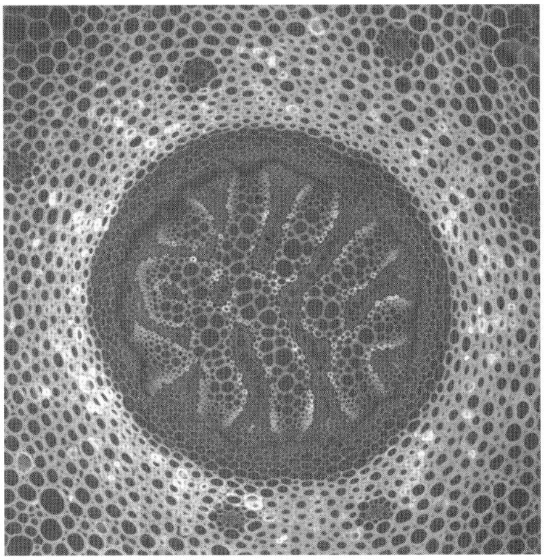

Figure 6.2. A cross section of *Lycopodium clavata* (club moss) showing that different types of cells are selectively stained by specific coloured dyes (Safranin O and haematoxylin). (Courtesy of Jim Haseloff, http://www.plantsci.cam.ac.uk/Haseloff)

(see Chapter 7 for further information on the developments of Ehrlich's ideas and the pharmaceutical industry).

It was to take decades before Ehrlich's ideas on selective toxicity were to influence the growing agrochemical and veterinary industries. By the 1930s, however, a number of companies soon found synthetic chemicals that would selectively kill plant pathogenic fungi or would control the insect pests of plants and animals. The 1930s was also the period of discovery of some extremely highly selective pharmaceutical drugs (see also Chapter 7). The synthetic sulphanilamide antibacterial substances came into use at around this time and a few years later penicillin (an NP) became available and was soon hailed as the first wonder drug. These advances began to reassure the public that a person could ingest a chemical without any expectation of an adverse effect on themselves but with a devastation effect on some organism within them. The public began to take the concept of selective toxicity to heart—clearly, not all chemicals were harmful.

The mid-twentieth century enthusiasm for synthetic chemicals

DDT

It was the selective toxicity of the insecticide DDT that was destined to have a most profound effect on attitudes to chemical safety. DDT was a chemical that had first been synthesised decades before the Swiss chemist Müller discovered its potent insecticidal action in the late 1930s. What was so remarkable about DDT was its selectivity. Even in extremely small doses, it was lethal to many species of insect yet it was remarkably non-toxic to humans even at quite high doses (Figure 6.3). The manufacture of DDT is

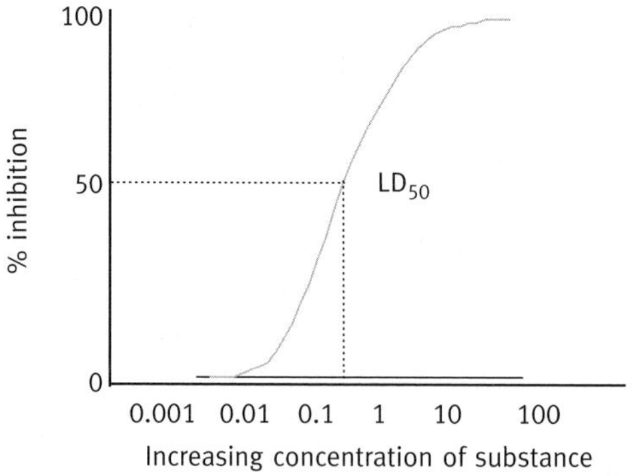

Figure 6.3. The LD_{50} is the concentration of a substance that kills 50% of a test population of organisms. It is a useful measure to compare the toxicity of different substances but it is a very controversial measurement when applied to mammals because of the suffering caused to the animals used to gain the data. Note that if the LD_{50} dose is increased by a factor of 10, nearly 100% of the population will die and if the concentration of the substance drops to one-tenth of the LD_{50} then only a very small percentage will die. The notorious insecticide DDT has an LD_{50} of 150–300 mg kg^{-1} for rats, 1 g kg^{-1} for goats, 2 g kg^{-1} for ducks but <0.1 mg kg^{-1} for some aquatic invertebrates.

relatively simple; hence, the chemical could be made cheaply in large quantities. Huge amounts of DDT were used in the Second World War to kill insect parasites (lice had been a very serious problem to soldiers and refugees during the First World War) and to control insect-borne diseases (malaria could debilitate an army). Such was the selectivity of DDT that refugees and soldiers could be casually dusted with DDT powder, during which time they would inevitably ingest particles. The persistence of DDT in clothing was regarded as a positive attribute because it gave lasting protection against re-infestation.

The impact of the introduction of DDT in reducing the death rate from malaria, caused by a mosquito-borne parasite, was especially dramatic. It has been claimed that by reducing the huge death tolls from malaria, DDT saved more lives than penicillin. DDT was cheap, effective and its persistence gave it a very long residual effect which allowed it to give a prolonged control of insects in houses when DDT was sprayed onto the walls of houses or even incorporated into wall paint. The attitude of most consumers to these new insecticides was transformed. In the early 1950s, the public were delighted to use DDT freely around their homes and gardens—the sticky fly paper in the kitchen was replaced by the insecticide sprayer in millions of homes.

DDT was not the only synthetic chemical to find a use in private and commercial gardens. Plant physiologists in the 1930s had accidentally discovered a way of selectively killing weeds in cereal crops (the dominant source of food for humans—wheat, barley, maize and rice). The plant physiologists interested in how plants controlled their growth had discovered a plant hormone—auxin (indole-3-acetic acid). Chemists soon found that, not only was it easy to make this compound, but they could also easily make hundreds of chemicals that were chemically related and structurally similar. A few of these *auxin analogues* were as biologically active as authentic auxin and these shared with auxin the remarkable property that when applied to a young cereal crop, the dose that had no effect on the cereal plants killed the majority of the broadleaved weeds. By the late 1940s, farmers in most developed countries were spraying their cereal crops in late spring every year with auxin analogues (the most widely used were 2,4-D and MCPA, both chemicals still being in use today). Weeds were not only a problem in cereal fields but also in grassland (cereals are grasses), so dairy farmers also began to use them. Given that grasses were also sown over large areas in urban settings (on golf courses, soccer pitches, cricket grounds, bowl greens, city parks and private gardens) these same weed killers were widely used in such settings. Families, and their pets, would use the lawns immediately after spraying without concern. New fungicides also became available to horticulturists and gardeners at this time and these 'safe' chemicals would be sprayed on garden flowers, vegetables or fruits in the garden, some of the latter would be consumed within days without concern.

Silent Spring

The public's benevolent attitude to chemicals was to change abruptly when the book *Silent Spring* was published in 1962. This remarkably effective polemic by Rachel Carson argued that the widespread use of pesticides was responsible for a dramatic decline in wildlife. The title of the book was a brilliant emotional hook for the public, despite the fact that most people rarely heard the dawn chorus or even knew it existed.[4] For the pesticide industry, *Silent Spring* was a public relations disaster. Birds have a special place in the minds of humans. Many humans feed birds in their garden or take their children to feed birds in a local park. In the United Kingdom, The Royal Society for the Protection of Birds (which has the largest subscription membership of any in the United Kingdom) has the title which suggests that birds deserve the protection of the monarch. One theme running through Rachel Carson's book was that it is the insidious nature of pesticides that allowed them to be widely adopted without early evident harm.[5] If prolonged exposure to small doses of synthetic chemicals could kill birds, maybe they could harm humans in the long term. The decline in iconic species like the birds of prey struck a special chord with the public. If the top predators were especially vulnerable, maybe humans, who are also near the top of the food chain, should also worry. A number of scientists had already identified some characteristics of one group of pesticides, the organochlorines (DDT, aldrin, dieldrin, etc.), that suggested that these chemicals might build up in tissues of non-target species after prolonged exposure to very low doses. Maybe the safety testing of new pesticides in the 1950s, testing which concentrated on the effects of acute (short-term) exposure, did not adequately reveal the chronic (long-term) effects? The organochlorines were characterised by high fat solubility which enabled them to penetrate the waxy insect cuticle. They also had great chemical stability which gave them the very long persistence much valued for some uses. However, these properties when combined also caused the organochlorines to accumulate in the fatty cells of any organism exposed to them and to persist for long periods. The result of these properties was that a small organism swimming in water containing a minute concentration of DDT would have a DDT concentration in its body that exceeded that in the water. A predator organism that consumed that little organism would accumulate the ingested DDT in its fat cells every time it fed. As one progressed up the food chain, the concentration of DDT was found to rise remarkably and the top predators would experience the highest doses. The public were soon alarmed that the same chemicals might be accumulating in humans.

Just at this time, the method of analysing samples for pesticide residues was revolutionised by the invention of the electron capture gas liquid chromatograph (invented by James Lovelock of the Gaia hypothesis fame). This instrument was especially good at detecting molecules containing chlorine; hence, the analysis of the organochlorines suddenly became something that could be carried out conveniently with huge sensitivity. Soon, evidence was published revealing the presence of organochlorine chemicals in samples of bird tissues, bird eggs, cattle, milk and even in humans milk—the latter hitting yet another button leading to human alarm.[6] The organochlorine contamination was found to be worldwide. Even organisms in remote ecosystems such as penguins were found to contain organochlorines. Despite the fact that there was no recorded human death from unintended DDT exposure, the seeds had been sown for distrust of pesticides in particular but synthetic chemicals in general.

Thalidomide

To compound this increasing scepticism about the use of synthetic chemicals, the pharmaceutical industry gave the world Thalidomide. In the late 1950s and early 1960s, Thalidomide was prescribed to large numbers of pregnant women in Europe, Canada and Australia to treat morning sickness. Despite very extensive safety testing of the drug before it was approved for use, it caused horrific deformities in a small proportion of the children born to women taking this drug. Inevitably, an increasing number of people began to note that scientists had been unable to protect wildlife, or indeed humans, from the adverse effect of synthetic chemicals. These problems had arisen despite the large number of scientific studies undertaken to address the issue of the safety of all such chemicals. The public scepticism of the skills and motivations of scientists grew.

The rising public concern

If scientists could be wrong about such widely used chemicals, and they could not foresee the very fundamental problems associated with particular chemicals, why should the public be reassured that other chemicals to which they were exposed were safe? Journalists had long known that scare stories help to sell newspapers; consequently, there would be few citizens in most developed countries that did not know of the debate about the safety of pesticides and pharmaceutical drugs by the 1970s. The concerns of the public were soon drawn upon by new political parties, such as the Greens, and some of the newly powerful Non-Governmental Organisations (NGOs). These groups soon became very effective lobbying organisations. Arguing that there would be a public benefit from a reduction in pesticide use, improved safety testing and more extensive environmental impact studies, Greenpeace, Friends of the Earth, WWF and their equivalents elsewhere gained wide support. These organisations were soon trusted more by large sections of the public than industry and government scientists. Indeed, the NGOs, which rose to power partly by questioning the safety of chemicals in the environment helped shape the wider political agenda in the last decades of the twentieth century so that environmentalism became part of the political and corporate mainstream.

The political, scientific and industrial responses

The sensitive antennae of politicians began to pick up signals that voters wanted a tighter regulation of the chemical, agrochemical and pharmaceutical industries. As usual when the public show concern about an issue, politicians are always ready to demonstrate their own concern by using public money to fund scientists to research the issues related to the problem. In the 1950s, it was toxicologists, studying the direct effects of chemicals on individuals in a laboratory, who had dominated chemical safety studies. Attention now turned to the effects, direct and indirect, of chemicals on populations. The growth of ecology as a subject at this time was greatly aided by new public funds directed to universities, government research institutes and government agencies to address these perceived problems with the use of chemicals. Inevitably, once these new academic disciplines had been established, the new breed of scientists dug deep to find problems as well as to suggest ways of resolving them. The funding of this research depended on the subject being kept 'live' and the popular media were ever ready to exploit any potential scare story that emerged from the scientific research.

Regulation

One of the most significant political responses to the public concern about chemicals, stimulated by Rachel Carson's book, was the establishment of the United States Environmental Protection Agency (EPA) in 1972. The EPA was very significant even beyond the shores of the United States. Because the United States was the largest single market in the world for the chemical industry, if a chemical manufacturer outside the United States wanted to access the huge US market, they had to comply with EPA demands. Furthermore, the agrochemical industry had already begun a process of consolidation in the 1960s and multinationals were growing rapidly. When US demands were made on the multinationals in order for them to sell into the US market, those demands would influence the multinational chemical and agrochemical companies worldwide. The EPA established a much more rigorous system of judging whether a chemical should be allowed on to the market (Table 6.1). Not only was much more data required for approval in terms of the toxicological properties of any individual chemical, but for the first time the environmental impact of the use of any chemical had to be fully considered.[8]

Table 6.1. In the early 1960s the increasing public concern about the safety of pesticides resulted in governments in the developed world requiring manufacturers to provide extra detailed results of more extensive safety testing of new pesticides.

	1950	1965	1980
Toxicology	Acute toxicity (LD_{50}) 30–90 d rat feeding	Acute toxicity (LD_{50}) 90 d rat feeding 90 d dog feeding 2 y rat feeding 1 y dog feeding	Acute toxicity (LD_{50}) 30–90 d rat feeding 90 d dog feeding 2 y rat feeding 2 y dog feeding Reproduction in 3 rat generations Teratogenesis in rodents Toxicity to fish Toxicity to shellfish Toxicity to birds
Metabolism	None	Rat	Rat and dog Plants
Residues	Food crops 1 ppm	Food crops 0.1 ppm Meat 0.1 ppm Milk 0.1 ppm	Food crops 0.01 ppm Meat 0.1 ppm Milk 0.005 ppm
Ecology	None	None	Environmental stability Environmental movement Accumulation Total effects on all non-target species

Source: Taken from Green, Hartley and West (1987).

Industrial consolidation

Throughout the last two decades of the twentieth century and the early years of the new millennium, industrial consolidation continued in both the agrochemical industry and the pharmaceutical industry. A number of the major agrochemical companies had diversified into the pharmaceutical market, often via explorations of the veterinary drug market.[9] However, in general, the exploitation of the principles of selective toxicity in the pharmaceutical industry was more profitable than the use of the same principles in the agrochemical industry. Consequently, after some mergers there was sometimes a realignment of the new multinational into two companies, one a pharmaceutical company and the other one an agrochemical company. It is possible that part of the motivation for this was to protect successful pharmaceutical brands from any adverse publicity that might come from the still existent adverse public attitudes to agrochemicals (and later GM crops because agrochemical companies had increasingly taken over the seed supply businesses).

The broader picture

Although public concern was focused on pesticides, and to a lesser extent on pharmaceutical products, some other types of products came to the attention of the public, hence alerting them to the fact that they were exposed to many chemicals.

Concern about food additives became so significant that food producers managed to seduce consumers with the prominent labels proclaiming 'free from artificial flavours' or 'no artificial colouring'. Such additives were blamed by some as being responsible for criminal, disruptive or violent behaviour of children. Food labelling, either under voluntary codes or by law, was introduced in response to such concerns. In Europe, the adoption of *E numbers* to identify commonly used food additives supposedly helped consumers make informed choices (or possibly helped alleviate some public fears).

By the end of the twentieth century, 30 years of increased safety testing, environmental monitoring and regulation had not entirely allayed public fears. The cost of some regulations was very considerable and in some instances these controls were based more on appeasing public concern than on scientific evidence.[10] For example, the European Union Drinking Water Directive addressed the issue of pesticide residues in drinking water. The European countries, along with many other countries, had previously enacted laws that imposed limits on the concentration of pesticides in drinking water, these limits being based on toxicological evidence. Choosing a figure for the limit for any chemical cannot be an exact science because the limit must always be set well below the concentration that clearly shows a statistically significant effect. This is a difficulty that few consumers and most politicians understand. The public and politicians want scientists to demonstrate that a chemical is safe—which is impossible because experiments can only show statistically meaningful positive effects. Practically, the best a scientist can do is to establish a concentration of a chemical that, under the conditions studied, will cause harm and then extrapolate from that information to predict a concentration that will cause no effect. Because the Law of Mass Action (as explained in Chapter 5) shows that the relationship between dose and response is log-linear (e.g., the dose of the applied chemicals has to increase by a factor of 10 to increase the effect by a factor of 2), a commonly accepted margin for safety is that it should be safe to consume 1/100 of the dose that is known from experimental studies to cause no measurable effect (Figure 6.3). However, such safety margins are extrapolations and are clearly based on many assumptions. Maybe humans are more sensitive than the species used to gain the toxicological data. Maybe laboratory bred strains of an organism show a narrower range of sensitivity than would be found in a human population. This is quite possible given that, in order to aid statistical inference, laboratory strains of organisms are often highly inbred, fed very uniform diets and kept free of disease and stress in order to produce a necessary uniformity. Maybe long-term exposure to low levels of the chemical will only reveal adverse effects after several years—humans live for decades, yet rodents, commonly used in laboratory studies, live for months. So, is the 100-fold safety margin sufficient? The response of the politicians to these dilemmas was not to side with the scientists who had spent decades trying to understand the possible effects of these chemicals. Instead, the European Union (EU) politicians decided that there should be *no* pesticides in drinking water. This simplistic idea, popular with voters, was clearly scientifically problematic, indeed it was unachievable. If one allows pesticides to be made and sold, they are inevitably going to enter drinking water. A farmer applying a fungicide to a field, a local authority using herbicides on footpaths, a timber treatment company treating woodworm in homes, indeed the gardener spraying their lawn or the pet owner throwing out their unused pet flea control agent are all opening up routes for their chemical to enter local water courses. Eventually, some of those chemicals will reach someone's drinking water, albeit in minute amounts. A very large number of studies have shown that chemicals are distributed globally by many different routes and even if one country tries to reduce the pesticide use within its borders, residues will still be found albeit at extremely low levels. Thus, the EU passed a law that aimed to achieve the impossible. However, another practical problem faced EU legislators, how could it be shown that a sample of drinking water contained no pesticide? Clearly by analysing water samples one can look for the presence of a chemical but how little of any pesticide can one detect? Analytical chemists advised that methods were available that could detect 0.1 $\mu g\ L^{-1}$ of any pesticide; hence, it was practical to use that concentration as the legal limit which must not be exceeded. So, we have a fine example of how politicians will happily ignore scientists when it suits them, yet at other times hide behind 'scientific opinion'. In this case, trying to achieve the impossible, EU legislators adopted a standard that was based on the wrong piece of science. The huge amounts of toxicological evidence and expertise that had been gained over the decades were simply put aside. The resulting cost of this political expediency was huge. In the UK alone, it was estimated that the water industry incurred capital costs in excess of £1.5 billion and annual running costs of £100 million to reduce the level of pesticides in drinking water from concentrations that were regarded by most experts as presenting no significant risk to levels that were even less of a risk. What the law did not achieve, and could not achieve, was to eliminate pesticides from drinking water![11]

The EU Drinking Water Directive has been discussed in such detail because it demonstrates how a few notorious examples of chemicals causing environmental and human welfare problems grew into such a general concern that legislators were ready to enact unsound laws, at huge expense to protect the public from risks which were minute compared to many other risks that the public accept. (The wily politicians also kept quiet about the fact that they have ignored their mantra of 'the polluter pays' because in this case the consumer was paying, not the farmers or agrochemical industry!)

NPs—what they can tell us about chemical pollution

Once the public had their attention drawn to the chemicals to which they were being exposed, it is hardly surprising that they found plenty to worry about. There are 75,000–90,000 synthetic chemicals in use, many of which have never been a subject of intensive toxicological testing. Even fewer have been subject to thorough environmental impact assessments. These facts were emphasised to the public, especially by some of the NGOs. Virtually, the only time members of the public heard or read about individual chemicals in the media was when they were mentioned as part of scare stories—pesticides in food, contaminants in tap and bottled water, side effects of drugs and so forth. Yet many people willingly expose themselves to hundreds of artificial chemicals throughout their day—starting with a highly scented, pigmented shower gel, hair conditioner, deodorant, cosmetics, dressing in polyester clothing recently washed in a detergent and softened with a fabric softener, drinking an artificially sweetened cola in a plastic bottle, chewing a fruit flavoured sweet, washing the crockery in detergent and so forth. Daily, hundreds of other synthetic chemicals are also likely to enter the body simply because it is inevitable that chemicals that are in use in society will always get into the environment; consequently, it is inevitable that they will be found in foods, water and homes.

It was inevitable that once the public had learned that chemicals could be dangerous, should anyone suggest that a specific chemical, or indeed even an unknown chemical, might be the cause of some newly perceived problem, citizens would look at chemicals as guilty until proved innocent. One might wonder why at the end of the twentieth century, when there are more people than ever who have had a scientific education, that the public has a sceptical attitude to chemicals. Given that the high standard of living enjoyed by most citizens of the developed world comes partly from the widespread use of chemicals, why is the public so distrustful of them? Why have scientists been unable to engage the public in a more helpful debate about the risk that chemicals in the environment pose to the average citizen? Scientists have not managed to give the average citizen enough understanding of the simple principles, outlined in the earlier chapters, to enable each citizen to make sound judgements about the risks that chemicals in their environment might pose. Indeed, so little confidence does the public have in scientific evaluations of chemical risks that the public are now quite prepared to ignore scientific evaluations on any topic—MMR, BSE, etc. Maybe if the arguments about chemicals in our environment were built on new foundations, the public might be better served.

Why an understanding of the evolution of NPs should underpin our attitudes to synthetic chemicals

The world has never been a chemically clean place

There seems to be an unstated assumption, held by many, that the world was a chemically clean place until humans began to manufacture chemicals industrially. This is completely wrong. Plants and microbes have been making maybe 500,000 different NPs for hundreds of millions of years and they continue to do so. Not only do organisms make a very large number of complex chemicals, they make them in very large amounts. It is hard to estimate how great this flow of carbon into NPs might be, but if only 1% of carbon fixed annually by plants (estimated to be 100 billion tonnes per annum) was converted into NPs, a total of in excess of a billion tonnes of NPs are made by organisms each year.[12] This figure suggests that biological chemical production greatly exceeds industrial chemical manufacture by a very large margin. Thus, humans did not begin something new when they started making and releasing chemicals, they simply contributed to a series of processes that have operated as part of the natural world throughout evolutionary time. Humans are now contributing very significantly to the quantity of chemicals being made annually and they have added to the overall chemical diversity but they are still minor players. Once one accepts the simple fact that humans are but the latest chemists in the world, the contributions that humans make to global chemistry and the way in which synthetic chemicals might have an effect on the world can be viewed with greater clarity.

Recognising that the world has contained hundreds of thousands of chemicals for billions of years, it becomes clear that most organisms must have evolved in a chemically complex environment. Some of the NPs are very potent poisons or have dramatic physiological effects on some organisms. Consequently, organisms will have evolved to cope with the chemical environment in which they have lived. It is predictable that organisms will have evolved mechanisms that enable them to survive, or thrive, in the presence of a mixture of chemicals, some of which may be harmful to them if accumulated or ingested in large enough doses. In the analogous way that individual organisms can survive very harsh physical environments, sometimes with remarkable morphological or behavioural adaptations, so it is to be expected that organisms will have evolved to cope with their own chemical environments. Just as most people from their own experience as animals can identify a few general principles that aid the survival of animals in harsh environments, so it is possible to identify a few general principles that will guide our consideration as to how organisms have evolved to cope with chemicals they encounter, willingly or unwillingly. It is to be expected that examples will be found where organisms

- avoid a particular chemical;
- are attracted to a particular chemical;
- possess mechanisms to reduce the concentration of a chemical in their bodies;
- possess mechanisms to enhance the retention of a particular chemical;
- have adapted to the presence of a particular chemical or group of chemicals; and
- have adapted to exploit a chemical that they do not make but have access to.

In other words, most organisms will have evolved to survive with chemical diversity in their environment. An organism exposed to a new natural or synthetic chemical will simply have one extra chemical in its environment. For reasons discussed in Chapter 5, the chances are extremely small that the new chemical will possess the particular properties that endow it with the potential to reduce the fitness of the organism. For billions of years, individuals of all species will, at intervals, have been exposed to a chemical that they have not encountered before. This will be a situation that might have happened many times in the lifetime of some individuals. It is certainly a circumstance that will have arisen many times in the recent evolutionary history of many species when species have increased their habitat range with the result that they inevitably encounter NPs that are novel to them. In other words, being exposed to new chemicals is a normal part of life. A good example of this fact comes from humans. Humans have been very successful at exploiting plants to add nutrition and interest to their diet. Every new plant added to the human diet contains hundred of NPs and most humans can ingest the new mixture of chemicals contained in novel food plants without ill effect.[13] So, how do organisms, including humans, cope with new chemicals in their environment or diet?

How have organisms evolved to live in the chemically complex world?

Avoiding a chemical

Location

Most organisms avoid most of the NPs in the world simply by living in particular places. Because most NPs are made only by a very limited number of species and because each species is usually only found in limited geographical area, the spatial distribution of each NP is usually very limited. The adaptation of any one species to a particular habitat limits the number of NP-producing species that will be contributing to any individual's chemical load. However, the more one species moves around, adapts to new locations and adopts new diets, the greater the range of NPs that species will encounter. Humans as a species must encounter many more NPs than most species (although goats have a formidable reputation as being able to eat anything).

Choice

Organisms that can move can also make behavioural choices to influence the type and magnitude of chemical exposure. The fact that many (most?) organisms, capable of movement, possess an ability to detect individual chemicals (using mechanisms analogous to those that humans call taste and smell), suggests that the ability to select or reject environments on the basis of chemical information, or to choose particular foods in that environment, was highly beneficial. Human experience shows that individuals make food choices on the basis of smell and taste, both largely governed by NPs. The co-evolution of insects and plants provides a further example of limitation of exposure by choice. Each specialist insect species limits its exposure to the number of NPs within its ecosystem and such insects have evolved in a chemical world that is less diverse than the chemical world of the local area, much less diverse than the chemical diversity of the country in which they reside and very, very much less than the chemical diversity of the world.

Reducing the concentration of a chemical

Even an organism reducing the chemical diversity and chemical load by selecting its food sources will still be exposed to some NPs. Without some mechanism to limit the concentration of these chemicals in the cells, the organisms might accumulate sufficiently high concentrations of some chemicals for those substances to have physiological effects on the organism and reduce its fitness. The Law of Mass Action (see Chapter 5) tells us that if an organism can keep the concentration of any potentially deleterious chemical below the threshold needed to cause an adverse effect, it will gain fitness. Surviving the ingestion of a toxic chemical is not about totally eliminating the potential toxin; it is about reducing the concentration of the substance to below a toxic level. The fitness gains of never needlessly eliminating a chemical will be significant. A mechanism that could potentially totally eliminate a specific chemical would not enhance fitness but might increase the costs; hence, such mechanisms would not be favoured by evolution.

Degradation and excretion

It is too simplistic to think that the metabolism of exogenous chemicals, whether natural or synthetic, by an organism is a way that the organism reduces the risk of exposure to an excess of any chemical. Consider the example of the human eating a pizza topped with tomato, green peppers, broccoli, mushrooms, capers and olives. Each bite will introduce a very complex mixture of NPs into the body—hundreds of exotic chemicals will enter the bloodstream. Each chemical will have properties that will govern its potential for harm. A very few of the chemicals might be quite toxic if given in larger doses but the concentrations achieved after eating the pizza will not be toxic—otherwise, humans would not eat them. A few of the chemicals might possess no potent biomolecular activity but might be converted by the pizza eater's degradative enzymes into new chemicals which possess much greater potential for harm. The unpredictability of the chemical mixture being encountered, the unpredictability of the properties of the chemicals being ingested, the unpredictability of the new chemical made as a result of the action of the degradative enzymes, these are all potential problems that the organism must have evolved to cope with. It seems unlikely that organisms with a varied diet, rich in NPs, will have evolved to possess enzymes targeted at each of the major NPs.[14] Given the degree of unpredictability that ingesting NPs brings, it seems possible that evolution has selected individuals that simply possesses the ability to keep the concentration of exogenous chemicals as low as possible, using generic mechanisms that are largely non-selective. Any organism with an excretion system possesses a route to dispose of chemicals; hence, it is predictable that mechanisms to direct ingested chemicals to that disposal route might have been selected. A plausible explanation is that organisms faced with a varying diet containing a mixture of compounds of unknown biological activity might be expected to have evolved mechanisms to rid the body of a wide range of substances irrespective of the biological or biomolecular activity of each substance. For example in mammals, most water-soluble chemicals ingested will automatically be diluted in the body simply because of the high water content of the body; furthermore, any soluble chemical will be excreted via urine. Consequently, there might be a rather limited selection pressure on organisms to evolve enzymes especially aimed at degrading most water-soluble NPs—dilution and loss via urine might be adequate to keep the concentrations of such NPs below the toxic threshold. More problematic would be highly fat-soluble NPs (or indeed any highly fat-soluble degradation products made by the organism in an attempt to degrade any ingested NP). Rather than being diluted throughout the body, fat-soluble NPs might accumulate in membranes or fat storage bodies and they could accumulate over long periods, eventually exceeding a toxic threshold. It is not unreasonable to look for an evolutionary solution to this problem that is generic rather than specific to each fat-soluble NP. Consequently, it is not surprising that there are some enzymes (e.g., cytochrome P450s) in the mammalian liver that act on molecules to add polar groups, such as hydroxy groups generating products that are more water soluble than the original NP. These enzymes typically have a broad substrate tolerance and so will act on many ingested compounds and will convert water-insoluble NPs to more water-soluble degradation products that will be diluted throughout the body and also excreted via urine. Thus, it is proposed that organisms will not have evolved 'to degrade toxins', rather they will have evolved to keep the concentration of exogenous chemicals below a toxic threshold. The very versatility of this system explains why most synthetic chemicals cause most organisms little harm—they have evolved mechanisms to cope with NPs and they are usually good enough.

This combined strategy of degradation and excretion would work by ensuring that there is a flow of chemicals from the body, thus keeping the concentration of any chemical at any potential active site low enough to reduce the chance of a significant interaction occurring. Such mechanisms would not be perfect but they would usually be sufficient, they would be versatile and robust if combined with other mechanisms such as learned behaviour. Individuals that persistently ingested high levels of NPs that overwhelm the mechanisms would be selected from the population and the individuals that favour a different diet would thrive. Human societies have culture as well as individual behaviour to guide individuals towards diets that are tolerated.

Isolation

All organisms can be regarded as possessing several 'compartments'. A cell is not a uniform entity rather it possesses a number of regions with specialised biochemical properties. These regions in eukaryotic cells can be surrounded by a membrane, thus they can possess some capacity to control their own environments. Once an organism is multicellular, the capacity for even greater spatial separation of functions becomes possible. The evolution of organs takes this specialisation a stage further. The ability of organisms to sequester (lock away) NPs in some limited regions of the organism might be another mechanism to avoid NPs accumulating in other more sensitive regions.

Adapting to a chemical

The past 100 years have shown us that organisms adapt to selective pressures imposed by the prolonged use of pesticides or drugs—organisms can adapt rapidly to new chemical selection agents. It is likely that such adaptation mimics the adaptation to NPs that has occurred in organisms throughout evolutionary time. An early example of organisms evolving the capacity to survive in the presence of a toxic chemical came shortly after the citrus growers in California had adopted the practice of exposing whole trees to hydrogen cyanide gas (by temporarily enclosing each tree in a portable 'tent') in order to kill insects overwintering on the tree. Within a few years of this practice being adopted around the beginning of the twentieth century, hydrogen cyanide no longer gave effective control of the insects, despite the fact that this gas is a very potent inhibitor of respiration. This surprising appearance of resistant organisms in response to intense selective pressure was to be the first of many examples that were reported during the twentieth century (Figure 6.4). Resistance of organisms to insecticides, fungicides, herbicides, some antimalarial drugs and antibiotics is now well known and a very serious problem. In insects exposed to unrelenting selective pressure, after 7–14 generations, it is usually possible to find mutants in a population that have developed some resistant to the normal dose being employed to control that organism. If the selection pressure is maintained, the descendants of the few surviving individuals will thrive and continued use of the insecticide at a higher dose will simply select for even higher resistance.

Now that the problem of resistance development is known, some strategies can be adopted which might reduce the rate of increase of resistance. For example, in some circumstances it is the exposure of some members of the population to sublethal doses for prolonged periods that increases the rate of development of resistance. Consequently, ensuring that the control agent is applied at the optimal concentration and for only a limited period can slow the rate of resistance development. Likewise, the use in sequence of two or more control agents, which operate by different mechanisms, can impose different selective forces in sequence and this normally reduces the rate of resistance development to either control agent. Furthermore, studies of antibiotic resistance in microbes suggest that although possessing the genes for resistance might itself impose additional costs on the resistant organism, subsequent mutations can reduce these extra costs so the resistance gene can persist in a population even after the use of the selective agent has been stopped. The management of the development of resistance is one of the biggest problems facing the agrochemical and pharmaceutical companies because if such resistance appears during the patent life of the control agent, profits might be very seriously reduced or even eliminated.[15]

Using the knowledge of the development of resistance to synthetic chemicals, one can predict that an organism encountering a new inhibitory NP in its food source, or in its environment, will evolve a capacity to adapt to the new chemical. By behavioural choices, or by chance circumstances, some individuals in the population will not be exposed to a lethal dose of the NP and the descendants of those individuals will eventually form resistant populations.

Exploiting a chemical

This topic is covered more fully in Chapter 8, where examples are given of NPs in the diet that are exploited rather than being degraded or excreted.

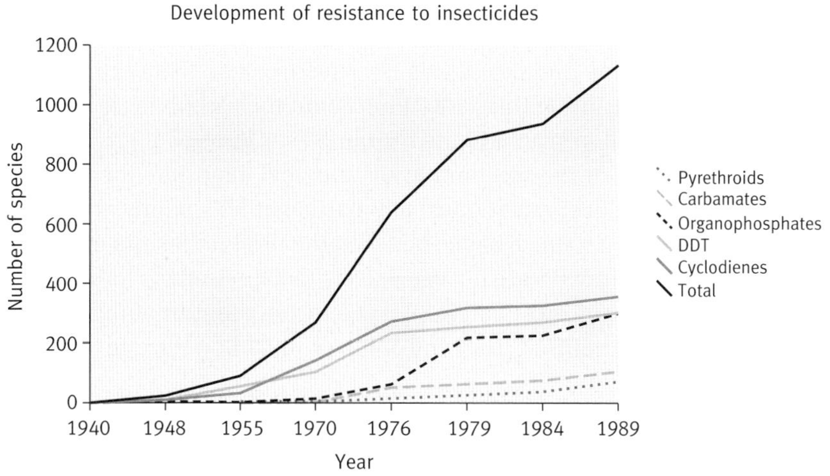

Figure 6.4. The evolution of resistance to toxin substances, as illustrated by insecticide resistance. As each new group of insecticide was widely adopted, resistance grew, but by the 1980s better management of pesticide use (e.g., Integrated Pest Management) helped to reduce the incidence of resistance for the newest insecticides (e.g., pyrethroids).

The breakdown of NPs and synthetic chemicals—why chemicals do not accumulate in the environment

The existence of NPs has primed the world for synthetic chemicals and mechanisms evolved in response to NP loads on organisms are being used by all organisms, including humans, to cope well with most synthetic molecules. This fact must not allow a complacency to develop about chemical pollution because, clearly, there are limits to the capacity of organisms to cope with chemicals in their environment, especially if those chemicals have some unusual properties (e.g., DDT). The analytical methods available to identify and quantify chemicals have advanced to the extent that few synthetic chemicals can hide from the determined investigator. However, analytical methods are always highly selective—they only find the substances sought or the chemicals with very similar properties to the one sought.[16] Modern analytical methods have been used to follow the fate of only a very few of the chemicals made by humans and which are released into the environment. Although we have a very incomplete record of the fate of the great majority of synthetic chemicals, there is little evidence that many such chemicals are building up in the environment. Why? Once again an answer to that question requires us to consider the fate of the older, more numerous, more chemically complex chemicals that have been released into the world for millions of years—the NPs. There appears to be some steady state of NPs in the environment and the mechanisms that are maintaining that steady state must be understood. So, how are the hundreds of millions of tonnes of NP lost from the world every year?

NP degradation—the NP cycle

If 1% of the carbon fixed by photosynthesis is converted into NPs each year, without degradation processes returning this carbon back into the carbon cycle, within a century the total world's carbon would be tied up in NPs. Although there is some accumulation of fixed carbon in the soil (and in oil and coal deposits), there is very little evidence for a significant accumulation of any particular NPs in any ecosystem. One must conclude that there is some balance between the production and of the degradation of NPs. Given that many NPs are not extensively metabolised by the organisms that make them, other organisms must carry out a significant amount of this degradation. There must be pathways that lead from any NP back to carbon dioxide. *There must be NP cycles.*

Microbes—an essential part of the NP cycle

The organisms that have most impressed humans by their ability to degrade synthetic chemicals are microbes (in the old terminology bacteria and fungi). Whenever attempts have been made to trace the degradation of synthetic chemicals in soils, microbes are found to be the main contributors to metabolic degradation. It is thus reasonable to postulate that these microbes are also the main contributors to NP degradation although much less attention has been paid to the microbial degradation of NPs.[17]

Why do microbes degrade synthetic molecules?

The existing paradigm

By considering the microbial degradation of both synthetic and natural chemicals as one, some interesting questions arise. Why do microbes carry out degradative processes? The conventional view, which is nearly universally accepted, is that microbes have a capacity to adapt to the availability of any chemical and that some mutants in a population will gain fitness by using specific chemicals as a source of elements or energy. The mutant that can use a novel chemical as a source of carbon, nitrogen or phosphorus to sustain its growth will be fitter than other individuals in the competing population. This view is part of the more general concept of an ecological niche, where every species has evolved to gain resources in a particular manner, balancing the ease of access to the resource with the ability to compete with other organisms. Becoming a specialist reduces the number and maybe the amount of available resources but also reduces the number of competitors for that resource. There is considerable evidence to support this model as applied to microbial communities, some of them are impressive. The most convincing experimental findings come from experiments where microbes are isolated and grown on simple mixtures of pure chemicals. Typically, a mixture of simple salts provide nitrogen, phosphorus, calcium and other minor elements and a single synthetic chemical provides the only carbon source for microbial growth. If any microbe can grow on this simple, well-defined culture medium, it must be able to access the carbon in the synthetic chemical; hence, the microbe must be able to break that chemical down into metabolisable substrates. The most well-known, simplest experimental method uses the algal polysaccharide agar to solidify the experimental solution on the base of a sterile Petri dish. A sterilised wire loop is used to smear a microbial sample (derived from a soil or water sample or from whatever source chosen for study) over the surface of the agar plate. After incubating the Petri dish for a period of days, it is usual to find just a few colonies of microbe growing on the provided mixture (Figure 6.5). From the millions of microbial cells on the agar plate, only a very few have an appropriate genetic makeup which enables them to degrade and utilise the synthetic chemical for growth.[18] This is the nearly universal finding, irrespective of the nature of the synthetic chemical incorporated into the agar or the source of the microbial cultures. The proportion of individual microbes that can access the substrate varies considerably, and the species that grow will vary but it is a rare chemical that will not be degraded by some microbe. What is really remarkable is that such simple methodologies readily find, *in most soil samples*, microbes that can utilise any one of the large number of synthetic chemicals. Sometimes, one can increase the success rate of finding an appropriate microbe to grow on a specific chemical by choosing a soil sample known to have been previously exposed to the chemical being studied. However, often a microbe that can degrade compound X can be found in a soil that has never been exposed to X. Why should natural microbial populations contain microbes that can utilise apparently unnatural substrates?

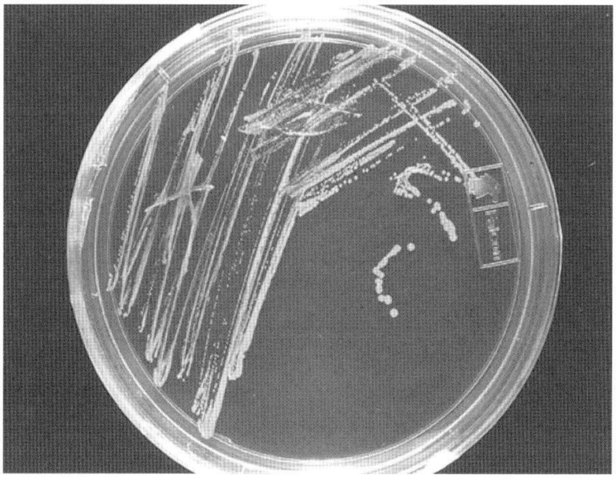

Figure 6.5. Selective isolation of microbes capable of living on a novel chemical.

The traditional view is that there are such a broad range of natural chemicals in the environment, chemicals which bear some structural similarity to synthetic chemicals, that there will always be some microbe that has evolved to specialise in degrading such carbon sources. The microbe that specialises in accessing a strange chemical, a substance that is not accessible to less specialised organisms, will have an enhanced fitness because it has exclusive access to that resource. If one considers each specific molecular structure as being equivalent to a niche, it is predicted that specialised microbes will evolve to exploit that niche. When a soil sample is cultured with only a single synthetic molecule as the carbon source, the only microbes that will grow will be those capable of accessing similar molecules in the soil.

The extensive experience of growing microbes on synthetic chemicals has been exploited in attempts to reduce the chemical contamination of soils. If microbes can be grown on a synthetic chemical and if it is possible to multiply them in the laboratory, one should then be able to apply such cultures to the soil to enhance the natural populations in contaminated soils. However, this approach has had less success than might have been predicted. Why?

Criticisms of the model

The paradigm very simply outlined earlier is an attractive one but one with some worrying features. One of the difficulties is that the selective conditions that apply on a Petri dish in a laboratory are very different from those that pertain in the natural environment. The selection pressure that operates when a microbial population is given access to only one novel substrate, a substrate that occurs at high concentration in the agar with all other elements conveniently at hand, is very extreme indeed. The scenario is hardly a model of the real world. In most soils, there are a very wide range of carbon substrates available, the most common being cell wall polysaccharides (such as cellulose) from plant material. If rare unusual substrates do occur in the soil, which they undoubtedly do (e.g., NPs from the roots, fungal NPs or microbial NPs or NPs from the decaying leaf litter introduced into the soil by worms or simply washed into the soil by rains), each one will occur at a concentration well below the concentration of the more common carbon substrates. Under these circumstances, organisms accessing the most accessible carbon substrates will be most able to compete for nitrogen, phosphorus and other elements. The organisms that can readily grow on the most widely available and accessible substrates would be expected to outcompete slower growing microbes struggling to access some poor quality carbon source that was only present at very low concentrations. Consequently, it is quite hard to explain why individual microbes with an ability to use an unusual, intractable substrate that only occurs at low concentrations would survive in soils. The circumstances that allow it to blossom when grown under the freak conditions of a selective culturing on agar are just too bizarre to be relevant to its survival in the real world. Indeed, if a microbe that can survive on a unique substrate is isolated and then that microbe is placed on a more accessible substrate, the ability to thrive on the intractable substrate is quickly lost from the population. It has been shown many times that if a microbe with an ability to degrade chemical X has been selected, if the microbe is grown on a more common carbon source such as glucose, mutants better suited to growing on glucose alone soon outcompete those able to use glucose and X. These findings explain why it is hard to use laboratory selected microbes to enrich the soil microflora in order to help breakdown some soil contaminant. Ignoring competition from existing soil microbes well adapted to that soil, mutants of the newly added microbe that gained the ability to grow on more accessible substrates that occur in the soil would soon appear and these would compete the strains originally selected.

So why do microbes possess enzymes capable of degrading synthetic chemicals

The alternative model

As noted earlier in this chapter, organisms making NPs have been releasing huge quantities of complex molecules into the environment for billions of years. Consequently, it is reasonable to deduce that the presence of NPs in the environment has been a factor in the selection of microbes with exotic chemical degrading capacities. However, in Chapter 5, it was explained that organisms making NPs might possess a repertoire of enzymes with low substrate specificity; consequently, an existing mechanism to generate chemical diversity could also be of value when there was a need to metabolise chemicals. Could it be that the broad substrate tolerance of enzymes making NPs make them prime candidates to produce the new degradative capacity? In other words, could the versatile, flexible metabolism of NP synthesis be the resource most likely to be drawn on for NP degradation? Maybe the difference between making and degrading complex substances is a blurred one in microbes? In this scenario, synthetic chemicals introduced into the environment are simply bringing about a quantitative change but little novelty is needed by existing microbes to cope with these new molecules—as explained in Chapter 4, there are few fundamental differences between synthetic and naturally made chemicals.

What does this chapter tell us about the way science works?

It was only when I was unexpectedly asked to take over a third year module on ecotoxicology, when a colleague was ill, that I began to reflect on the way in which my knowledge of NPs might inform the new subject I was learning. Nowhere in my reading on pollution and toxicology did NPs appear but I did begin to realise that many books on pollution or ecotoxicology listed a rather small number of examples of chemicals that were a cause of concern. One book listed 80 chemicals in its index yet I had read elsewhere that tens of thousands of synthetic chemicals had been made industrially and I knew that estimates of the number of structures made naturally ran into the hundreds of thousands. Being an avid reader of Arthur Conan Doyle, I recalled Holmes's deduction, given in the quote at the beginning of this chapter, that the lack of something happening was often highly significant. It is surely interesting and significant that tens of thousands of synthetic molecules seem not to have caused environmental concern and less than 100 are perceived as a problem. Instead of thinking about those 100, why not think about as to why the great majority were not a problem? Given that I had thought a lot about whether synthetic and natural chemicals differed, it suddenly occurred to me that there were no known problems caused by NP accumulation and that was also a significant fact. A consequence of putting these ideas together produced this, no doubt controversial chapter. However, what I think the chapter shows once again is that scientific subjects can become too closed and insular. The practitioners can become too comfortable within the parameters they (or their teachers) have set. All scientific knowledge needs to be joined together, yet so often we teach it as if each topic self-contained.

The way in which politics interacts with science is well illustrated by the EU Drinking Water Directive story. Politicians have been happy to follow scientific advice when it suits them and ignore it when such advice is politically inconvenient. One problem is that politicians, like many of the public, do not recognise that scientific advice is never going to be unambiguous because all scientists can do is to form an opinion about evidence, sometimes using a good theoretical understanding to make sense of the evidence but sometimes using a defective theoretical model to shape their thoughts. History should tell us that one generation's orthodox scientific ideas have a habit of being overturned or greatly modified by the next generation. Our generation would be very arrogant indeed if they assumed that they are the first generation to have only the right ideas. Scientist must argue and disagree, that is their job. History tells us that the view held by the majority at any one time is not always correct; hence, the scientific advice to politicians can never be perfect, especially if it comes from one committee chaired by a single strong personality.[19] Society would be better served if at least two independent committees could ponder any scientific issue of public concern and argue about their independent conclusions in an open, positive way. In other words, the model used to incorporate scientific knowledge into public policy badly needs reform. Why otherwise would we still be doing so little about the potentially disastrous consequence of global warming yet still spending large sums to clean drinking water for no scientifically valid reason?

7

Natural Products and the Pharmaceutical Industry

A drug is that substance which, when injected into a rat, will produce a scientific report.

—*Author unknown*

Summary

Some NPs have been exploited by humans as a means of treating illness, ailments and infections throughout recorded human history. Worldwide, NPs are still of major importance in health care. However, there is currently a great debate as to whether NPs will retain their importance as pharmaceutical agents. By the end of the twentieth century, all the major pharmaceutical multinational companies had massively reduced investment in NP research. Concurrently, many people interested in international development argued that it was time to increase the study of NP diversity in the less developed countries (bioprospecting) in order to provide novel drugs for the rich and to provide an income stream for the poor. So, where should investments be made in order to increase the chances of finding important new pharmaceutical products? Applying the scientific principles outlined in earlier chapters, it can be shown that an understanding of the evolution of NPs could help transform our approach to finding and exploiting novel pharmaceuticals.

Herbalism

Throughout history, people from all cultures have used herbs and plant extracts to treat illnesses and to dress wounds. There are reports of chimpanzees choosing to eat the leaves of some species that may reduce their gut parasite population. Cats are regularly seen to chew the leaves of grasses, yet ignore other plants (although some cats do get excited by the NPs in catnip). Because of the very powerful placebo effect associated with taste and smells, and because many of the major groups of NPs have strong odours or flavours (see Chapter 8), it is possible that plants and fungi rich in NPs were appealing to some species of animals seeking to improve their health. The increasing intelligence of hominids would have enabled those species to make clearer associations between specific ailments and their treatment with NP-containing plants and fungi. Certainly, as soon as recorded history begins, herbalism is clearly well established. We have thousands of years of evidence of the practice of herbalism in many cultures, some hundreds of years of quite detailed treatises on the subject and even in the twenty-first century the majority of the world's population rely heavily on plant extracts containing NPs, or purified NPs, for the treatment of ill health. Indeed, it was only in the twentieth century that the monopoly of NPs as pharmaceutical products was challenged by synthetic chemicals. Over four billion people currently use herbal medicines and even in advanced industrial countries NP-based medicines are widely used. Nearly 60% of Germans buy such medicines and the sales of herbal medicines in the United States were $85 billion in 2007.

The modern pharmaceutical industry starts with a search for a way of making an NP

As described in Chapter 2, in the mid-nineteenth century, the need for greater and more secure supplies of quinine to treat malaria was a challenge that the increasingly confident synthetic chemists were ready to accept. The great race among the newly industrialised European nations for African colonies had begun,[1] and explorers, settlers and the military all needed quinine to enable them to survive infection with the malaria parasite. Although a route to synthetic quinine eluded chemists at that time, Perkin's attempt accidentally led to the discovery of the dye mauve and the birth of the hugely successful dye industry. The growth, and economic importance of the German dye industry in particular, was a major stimulus to the blossoming subject of synthetic chemistry in the late nineteenth century. A very large number of synthetic dyes of all shades and hues were developed and this allowed fashionable colours to change with the seasons—a dominant feature of fashion that remains to this day. The chemical stability and photostability (resistance to fading in sunlight) of the synthetic dyes was essential for their use and some were much more stable than natural vegetable dyes (for reasons discussed in Chapter 4). It was one of these stable dyes, methylene blue, that was to be of particular significance in the establishment of the modern pharmaceutical industry. A young German scientist, Paul Ehrlich, was given the task by the great chemist Hoffman[2] of trying to establish the path of infection of malaria. Ehrlich found that the methylene blue staining of the parasite in an infected sailor allowed him to trace the protozoan. The parasitic cell had taken up the dye to such an extent that it became visibly stained against a background of cells that were not stained. This suggested that the concentration of dye in cells depended on the cell type. Furthermore, the sailor seemed to recover and showed no ill effects from his exposure to methylene blue. Ehrlich deduced that when many different types of cells were exposed to the same concentration of methylene blue, one cell type must be receiving a high dose of the chemical while another cell type must receive a much lower dose. Given that it had been appreciated for many centuries[3] that poisons only acted when they were administered above a certain dose, it was reasonable to think that it might be possible to give high (poisonous) doses of a chemical to some cells while leaving other cells unharmed. The idea of *selective toxicity* was now based on experimental findings, and Ehrlich was to develop his career with this concept as a focus. It was an exciting time for medical research because it had been established that infectious diseases were caused by an invasion of the body by simple organisms such as bacteria or protozoa. Some of these organisms could be cultured in the laboratory or removed from infected animals and subjected to microscopic study. The small size of these infectious organisms facilitated simple laboratory studies to test whether a particular substance was toxic to the organism. It was practical to 'screen'[4] collections of chemicals to find the few chemicals that produced the desired effect at a low dose. The combination of synthetic chemists producing thousands of new chemicals and biologists devising practical ways of testing, cheaply and rapidly, each chemical for some specific biological action was at the heart of the new pharmaceutical industry. However, as discussed in Chapter 5, it was soon realised that a truly selective toxic agent was a very rare chemical indeed. Furthermore, a promising chemical found in a laboratory screening trial often had undesirable side effects when given to an animal. These two problems were to hinder the pharmaceutical industry for the next century, indeed they still hinder it. However, the rewards from finding the very rare truly selective agent were so great that investors found the returns worthwhile and major pharmaceutical companies blossomed in several European countries, in the United States and Japan. Ironically, although Ehrlich started his work studying malaria, a successful treatment for malaria eluded him, and many others, and it took his successors in Germany until 1926 to finally discover the effective synthetic antimalarial drug pamaquine. However, Ehrlich was more successful in finding drugs to treat other infections. In 1905, he showed that trypan red, another dye, was effective at treating sleeping sickness, an infection that debilitated many in the new European colonies. In 1910, Ehrlich discovered Salvarsan as a treatment for syphilis, producing a treatment of interest worldwide to all sections of society.[5]

Throughout the developed world and throughout the twentieth century, companies were attracted to the huge profits that were available to those who held a patent on a successful pharmaceutical or veterinary agent. In the first half of the twentieth century, the cost of entry into the pharmaceutical or veterinary market was not very high—some simple laboratory facilities were all that were required for the discovery process. Consequently, nearly all countries with well-developed academic institutions teaching chemistry, medicine and biological subjects spawned small pharmaceutical companies. Few such companies could challenge the dominant German, UK, Japanese, Swiss and US companies but many survived making licensed products, or products out of patent, for local markets. However, as drugs were more widely sold and as medical knowledge increased, it became apparent that even 'selective' agents were rarely completely selective. Even if only a small percentage of the population treated showed side effects, drugs could be devastating to individuals. To guard

Table 7.1. The largest pharmaceutical companies and their research and development budgets in 2003. By 2008, the largest Pfizer, had increased their sales to $44,000 million, a sum that exceeds the GDP of over 50% of the world's economies. Four of the companies in the list have merged with others in the subsequent five years.

Company	Annual sales (million $)	Annual R & D spend(million $)
Pfizer	28,288	5,176
GlaxoSmithKline	27,060	4,108
Merck	20,130	3,957
Astra Zeneca	17,841	3,069
Johnson and Johnson	17,151	3,235
Aventis	16,639	2,799
Bristol-Myers Squibb	14,705	2,746
Novartis	13,547	2,677
Pharmacia	12,037	2,218
Wyeth	10,899	2,359
Lilly	10,285	2,080
Abbott	9,700	2,149
Roche	9,355	1,562
Schering-Plough	8,745	1,425
Takeda	7,031	1,304
Sanofi	7,045	1,152
Boehringer Ingelheim	5,369	1,020
Bayer	4,509	1,014
Schering AG	3,074	896

against highly expensive litigation, pharmaceutical companies had to undertake much more extensive safety testing of newly discovered drugs. Governments also demanded even more data before allowing a drug to be sold commercially. The demands made on companies to gather data to show that drug treatments were both safe and effective massively increased the cost of drug development. These extra costs increased the cost of entry into the industry and forced many smaller companies to merge with other companies. The expensive marketing of branded drugs also became increasingly important. Consequently, the latter half of the twentieth century saw a gradual consolidation of national pharmaceutical companies and eventually a more rapid consolidation into the giant multinational pharmaceutical companies that dominate the industry today (Table 7.1). The largest of these pharma[6] companies are among the industrial giants of the world economy. The combined sales of the pharma companies make this industry the largest legal human activity with current annual sales exceeding $400 billion.

NPs in the pharmaceutical industry—the era of antibiotics

The market for pharmaceutical products is one that will continue to grow because only a small proportion of the world's population currently has access to the most modern drugs and as people live longer they need more medical interventions. Significant proportions of pharmaceuticals, or the precursor chemicals used to make them, are NPs from plants or microbes. Some estimates suggest that over 25% of the drugs sold in the developed world and 75% in the low-income countries (LDCs) are based on NPs.[7] Why after more than a century of intensive efforts to make synthetic drugs are NPs still so important? After Ehrlich's success in finding new drugs among the growing collection of synthetic chemicals, it looked for a few decades as if NPs would be eclipsed by synthetic drugs. However, a new golden era was about to begin for NPs, and that era began with successful introduction of penicillin as an antibiotic. Such was the power and selectivity of penicillin that a massive hunt for new microbial NPs began. These searches once again placed NPs back at the centre of drug discovery programmes from the 1950s until the 1970s. Spurred by the dramatic success of penicillin, nearly every large pharmaceutical company in the world started a microbial screening programme in the hope, and expectation, of finding a novel antibiotic. The underlying logic was really economic but there was a scientific justification that could be used to convince any sceptical shareholders. If, as was increasingly believed by many scientists, microbes made antibiotics in order to defend themselves against other microbes, there must be many new antibiotics awaiting discovery; the first to find them could patent them and make a fortune.

Penicillin

The 1945 Nobel Prize for Physiology or Medicine for the discovery of penicillin was shared between the microbiologist Alexander Fleming (who worked at St Mary's Hospital in London), Howard Walter Florey (Professor of Pathology at the University of Oxford) and biochemist Ernst Boris Chain (a member of Florey's team at Oxford University). The story of the discovery of penicillin is as complicated as the characters involved. Most accounts begin in 1928 with the Scot Alexander Fleming (1881–1955) finding by chance that the blue-green mould *Penicillium notatum* secreted a substance that inhibited an adjacent colony of *Staphylococcus aureus*.[8] However, it is now agreed that the young French medical student Ernest Duchesne has a prior claim as the discoverer. In his 1897 dissertation, Duchesne reported that a *Penicillium* mould contained a potent antibacterial substance. Duchesne partially purified the antibiotic and even carried out a successful assessment of the antibiotic properties of the extract in animals. Unfortunately, Duchesne died at an early age in 1912, but it now appears that Fleming was really rediscovering something that had already been found, even if it was not widely known. In 1929, Fleming published the results of his investigations in the *British Journal of Experimental Pathology*, but he never succeeded in producing enough of the active substance to follow up his early observations and he turned his attention to other lines of study. Significantly, at about that time the first of the effective synthetic antibacterial compounds were exciting interest. The antimicrobial sulfanilamide drug Prontosil was shown in 1935 by G Domagk to be converted in the body to an analogue of the vitamin *p*-aminobenzoic acid and he was awarded a Nobel prize for demonstrating its effectiveness against *Streptococcus* and a broad range of other microbes.[9] However, in 1935, the Australian Howard Florey was assembling a team of researchers in the pathology department at Oxford and among those he recruited was the volatile, talented European refugee Ernst Chain. Although notionally recruited to work on cancer, Chain had an interest in Florey's work on the ability of lysozyme (the enzyme found in tears) to kill some bacteria by lysis (breakdown). Chain began reading more about antibiotics and in 1938 he read Fleming's 1929 paper and it fired his imagination. He repeated Fleming's observations and soon made more progress than Fleming had done. Recognising the significance of the work and stimulated by the thought that in the expected Second World War a large numbers of troops would, as in the First World War, die of bacterial infections, Florey secured government funds to investigate the possibility that Fleming's substance could be a useful antibiotic. Recruiting the modest, but technically imaginative and ever resourceful Norman Heatley to the team, Florey began to culture the mould in increasing quantities despite the limitations due to the outbreak of war. Enough of the substance, soon to be called penicillin, was isolated and partially purified to enable a trial to be made of its effectiveness as an antibiotic on infected mice. Not only did the penicillin cure the infection but the mice showed no significant side effects of the treatment. However, despite Heatley's best efforts, using bed pans among other containers to grow the mould, the production of penicillin was very limited. At best, only a few milligrams of penicillin per litre of medium was produced, so the typical current dose would have required 200–2000 litres cultured media using Heatley's methods. To compensate, when sufficient penicillin had been accumulated to try on the first patient in the Oxford's John Radcliffe hospital, the patient's urine was re-extracted to glean extra supplies of penicillin to continue the treatment. However, the effectiveness of the antibiotic was so impressive that it was clearly a matter of urgency to find ways of increasing production. The chemical structure of penicillin (Figure 7.1) was being studied to ascertain whether it would be possible to make the chemical synthetically. When the structure was established, the 1000 chemists set to the task of finding a synthetic route were unable to achieve that goal. Fortunately, the ever resourceful Heatley and the determined Florey were sure that the yields of penicillin could be increased from *Penicillium* cultures and improved methods of isolating the substance could make penicillin a practical treatment. Florey decided that a major effort of research and development was urgently needed and he contacted some drug companies in the United Kingdom, and via intermediaries, some of the large US drug companies. Florey had met and made an ally of Lord Rotheschild and through him he met an official in the US drug agency who became Florey's champion in the United States. In July 1941, it was arranged for Florey and Heatley to visit the United States to meet some representatives of the large US drug companies. However, Florey was disappointed and frustrated when he received a rather lukewarm reception from some of these companies. But slowly, and later more rapidly, as the potential of penicillin became clearer to sceptics, a programme was established in some companies to devise methods of mass production of penicillin. One breakthrough came at an

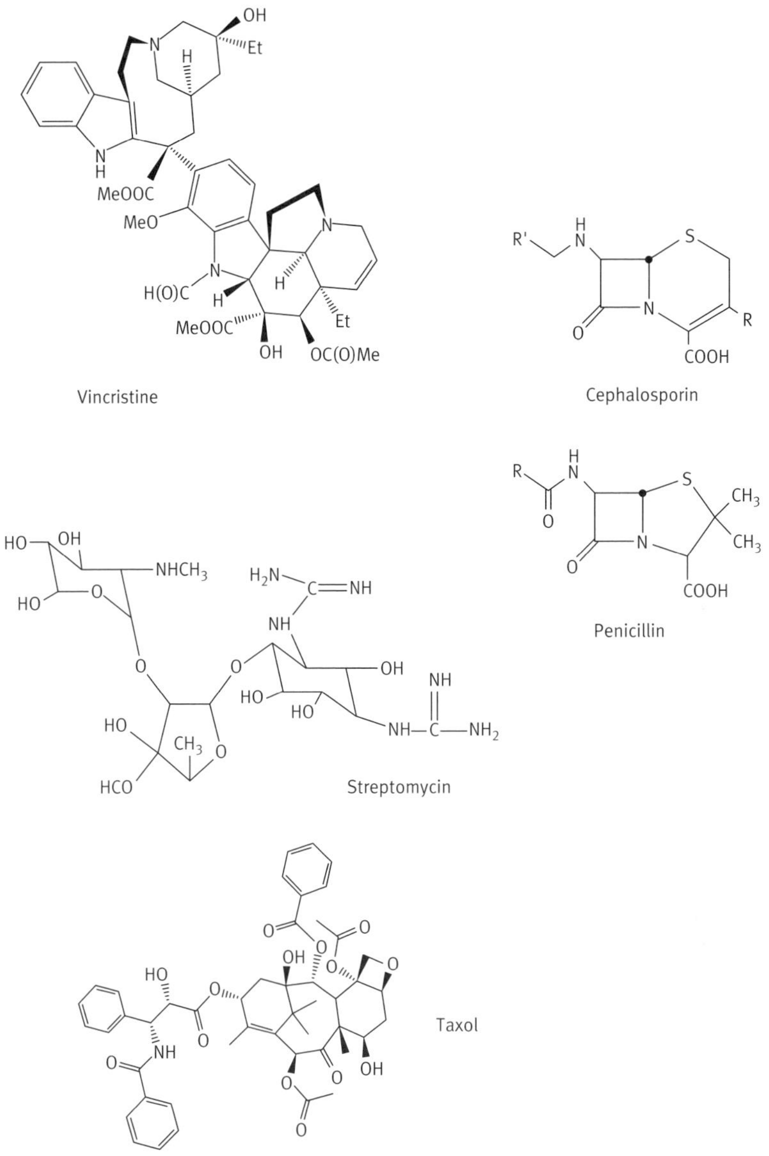

Figure 7.1. The structures of the major NPs of importance as pharmaceutical drugs.

unlikely venue. The Peoria Laboratory in the US midwest had been given the task of finding ways of helping the agricultural economy in the area and was looking for some way of using the huge amounts of corn steep liquor that was a waste product of corn starch production. They had found that it was a very useful fermentation medium and had devised ways of culturing microbes in massive airlift fermentors. Heatley joined the Peoria Laboratory for some months, where he shared his experience with the local experts and soon the group greatly increased the yield of penicillin in the cultures. Part of the project also involved screening new samples of *Penicillium* mould to see whether they could find an isolate which was inherently more efficient at penicillin production. The US military, now establishing new bases worldwide as part of the war effort, were asked to send soil and vegetation samples to the Peoria Laboratory and thousands of samples were evaluated without significant success. Ironically, a technician in the laboratory spotted a nice green mould on a melon in a local Peoria market and when they cultured that local sample it was the most effective penicillin producer of all the samples tested. This local mould became the source of the penicillin as it went into production. By late November 1941, Andrew J Moyer, a Peoria expert on the nutrition of moulds, and Norman Heatley had succeeded in increasing the yields of penicillin 10-fold. More extensive, highly successful clinical trials took place in 1943, and penicillin production was then rapidly scaled up so that supplies were available to treat Allied soldiers wounded on D-Day. The improved production and isolation methods allowed the price to drop from $20 per dose in July 1943 to $0.55 per dose by 1946, a quite remarkable achievement. Current fermentation methods and high-producing strains now make it possible to produce 1000 times the amounts that Heatley could make.

Although penicillin was never patented, a fact that caused some friction between Florey and Chain, patents were granted on some of the improved methods of production developed by the Peoria Laboratory and by some of the industrial laboratories that had also become interested. For example, in 1948, Andrew J Morton was granted a patent for a method of the mass production of penicillin.

The discovery of penicillin placed NPs back on the agenda of all the major pharma companies. Improved methods of production were developed and chemically modified penicillin analogues, with improved clinical value, were patented and widely adopted. Such was the optimism engendered by penicillin that it was rashly predicted that bacterial diseases would eventually be eradicated from the human population. Anyone with a reasonable knowledge of evolution and of NPs would have been surprised had that prediction come true.

Why devote so much space to the discovery of penicillin? Simply because penicillin was the first NP to be made in massive amounts in factory scale fermentations, because of its remarkable biomolecular properties. This showed, for the first time, that microbially produced NPs were economically accessible to large populations of humans and that chemists had no monopoly on synthetic methods for the pharmaceutical industry. The story also tells us that a worldwide search for cultures best suited to making penicillin showed that it is the rare organism that makes antibiotics in large amounts, a conclusion confirmed by the next part of the story of antibiotics.

Streptomycin

In the late 1930s, another search was underway for microbially derived antibiotics, a quest lead by the Ukrainian immigrant to the United States, soil microbiologist Selman

Waksman of the Department of Soil Chemistry and Bacteriology at Rutgers University in New York. This was a planned programme of screening that eventually led to the discovery of streptomycin from *Streptomyces griseus*. Streptomycin was active against a number of bacterial diseases and was especially valuable because it was active against some species that were not controlled by penicillin. Streptomycin was effective against tuberculosis (*Mycobacterium tuberculosis*), walking pneumonia (*Klebsiella pneumoniae*), fowl typhoid (*Shigella gallinarum*), one of the bacteria involved in some food poisonings (*Salmonella scottmuleri*) and two bacteria that cause urinary infections (*Brucella abortus* and *Proteus vulgaris*). Selman Waksman was awarded the Nobel Prize for Physiology and Medicine in 1952 for his 'ingenious, systematic and successful studies of soil microbes that have led to the discovery of streptomycin, the first antibiotic remedy against tuberculosis'.[10]

Streptomycin is one member of the family of aminoglycoside antibiotics. Members of the family made by strains of *Streptomyces* have names ending with *-mycin* and those made by cultured strains of *Micromonospora* have names ending with *-micin*. These NPs inhibit protein synthesis in various types of bacteria. Unfortunately, some of the family have adverse effects on kidney functioning or hearing in the treated patients; hence these drugs tend to be used as a second line of defence. Bacterial resistance has also become widespread, with several predictable mechanisms recorded. Some resistant strains have evolved with changed mutated proteins on the 30S ribosomes, proteins that bind the antibiotic less strongly. Other strains take up the antibiotic poorly and some can degrade the antibiotic. The latter form of resistance is due to the production of an enzyme coded for by extrachromosomal DNA that is carried by a plasmid (a small circular piece of DNA that can be passed between bacterial species). This typical example of detailed investigations of the cause of antibiotic resistance development helps to form ideas about the role of antibiotics in evolution, which are discussed in the next chapter.

Gramicidin

In 1939, René Dubos, Waksman's former postdoctoral student, extracted two chemicals, tyrocidine and gramicidin, from the soil germ *Bacillus brevis*. These chemicals cured bacterial infections in cattle but were too toxic for humans. This discovery prompted a number of scientists to expand the search for microbes in the soil, microbes capable of making chemicals that could kill disease-causing bacteria in humans.

NPs in the pharmaceutical industry—the synthetic steroids

Synthetic steroids

By the 1930s, it was clear that some humans suffered from deficiencies of steroids, compounds related to the steroidal NPs found in plants and microbes (see Chapter 9 for the debate about whether some steroids should be classed as NPs). The use of steroids to treat such patients was limited by the supply of steroids which had to be laboriously extracted from animal-derived material and, consequently, were prohibitively expensive. The US chemist Russell Marcker, working at Pennsylvania State University, realised that it should be possible to make human steroids from the diosgenin, the structurally related steroidal compound made by plants. This did indeed prove feasible when it was shown that a steroidal extract of a Mexican plant could be converted to the human female hormone progesterone. Unable to gain interest or support from the US drug companies for his discovery, Marcker found a Mexican businessman ready to invest in a new company, Syntex (*Syn*thesis + *Mex*ico), to exploit this discovery. Sadly, Marcker was swindled of his share of the profits that soon flowed from this company. When he tried to form a rival company, he was subject to physical and legal harassment and maybe wisely retired from industrial chemistry in 1949 and became a dealer in Mexican antiques. However, Marcker had begun an industry that blossomed in subsequent decades as the contraceptive pill, based on plant-derived synthetic steroids, became a major pharmaceutical product and helped women in many countries, both developed and developing, make their own reproductive choices for the first time.

The next great chemist to take up the challenge of making other human steroids was the Austrian-born Carl Djerassi, who fled his country after the Nazi invasion in 1938, and joined the Swiss owned CIBA company in New York. He subsequently joined Syntex (now a respectable company after that shady start) and devised a way of making cortisone from extracts of Mexican yams or sisal. However, the Syntex synthesis of cortisone was never commercially successful because a competing method, involving the use of microbial fermentation, could provide a cheaper product.

NPs in the pharmaceutical industry—the era of anticancer drugs

Vinblastine

One of the most valuable treatments of several forms of leukaemia is the NP vinblastine. Its discovery is yet another example of serendipity. In 1952, the Canadian Dr Robert L Noble (Associate Director of the Collip Medical Research Laboratory at the University of Western Ontario) received an envelope from his brother Dr Clark Noble containing 25 leaves from the Madagascar periwinkle plant (*Catharanthus roseus*). One of Clark Noble's patients in Jamaica had told the doctor that a periwinkle tea was used in Jamaica for diabetes treatment. Dr Robert Noble started an investigation of the properties of extracts of the leaves but he found that there was little effect on blood sugar levels but unexpectedly white blood cell counts had decreased in animals treated with the extract. Given the fact that the uncontrolled production of white blood cells was associated with leukaemia, this finding suggested that a periwinkle leaf extract might be worth investigating as a treatment for leukaemia. In 1954, Dr CT Beer (an Oxford trained organic chemist) joined Dr Noble's research team[11] and by 1958 they had successful isolated and purified a potent alkaloid extract from the leaves. They named this extract vinblastine. Collaboration with the pharmaceutical company Eli Lilly followed and sufficient vinblastine was produced for clinical trials to begin in 1959 at the Princess Margaret Hospital in Toronto. While not a cure, vinblastine in combination with other drugs was very effective in controlling the growth of a number of different types of cancers. Vinblastine is still one of the most useful chemotherapeutic agents available and its discovery and isolation is considered to be a milestone in the history of cancer chemotherapy, particularly for the management of Hodgkin's disease and testicular cancer.

Vincristine

Given Eli Lilly's involvement in the development of vinblastine, it is not surprising that the company funded a team to investigate the other alkaloids in *Catharanthus roseus*. One of the most potent alkaloids was given the name vincristine and approved for drug use in 1963, initially as a treatment of leukaemia. Vincristine acts by binding to the microtubules in the cell, disrupting, among other things, cell division.

Eli Lily currently have vinblastine and vincristine sales that exceed $180 million per annum.

Taxol

Because of the success in the 1940s and 1950s of finding the major pharmaceutical agents described above, in 1958 the US National Cancer Institute began possibly the world's largest NP screening programme ever, seeking a chemical that might usefully treat some form of cancer. The selective toxicity sought would have to be extreme because the cells that were the targets were not those of another species but abnormal human cells. The programme to screen hundreds of thousands of extracts containing NPs met with very little success. However, in 1963, the US Forest Service provided a sample of Pacific Yew tree (*Taxus brevifolia*) for extraction and testing. Unlike most plant samples tested in the programme to date, extracts of this tree were found to inhibit cell division. However, progress was slow and it took until 1971 before the compound responsible for the activity, named taxol, was identified and characterised. Taxol, like several other anticancer drugs, binds to microtubules, consequently interfering with cell division. The structural complexity of taxol suggested that a chemical synthesis would be extremely challenging and the difficulty in obtaining sufficient material from the forest trees suggested that there might be no commercial future for the product. However, small-scale studies continued and there was a renewed interest and excitement in 1989 when some women suffering from ovarian cancer responded very well to taxol treatment. These results changed the outlook for taxol. An agreement was signed with the large pharmaceutical company Bristol-Myers Squibb for further development and marketing of taxol and a large investment was made both in the possibility of chemically synthesising the drug or finding better natural sources. The challenge of producing enough taxol by extracting plants was a huge logistical task. Like many NPs, the concentration of taxol in the plant is very small; the bark of Pacific Yew trees contains only 0.02%. Furthermore, the removal of the bark for extraction kills the tree. In order to produce 1 kg of taxol, 3000 trees have to be sacrificed. It was calculated that to treat ovarian cancer with taxol in the United States alone would require the destruction of 75,000 trees per year. If the drug were to be made available worldwide, hundreds of thousands of trees would need to be harvested annually. This alarmed conservationists. The harvesting of the Pacific Yews would inevitably lead to the destruction of other trees and the habitat would be degraded. The forests in question were the home of the Spotted Owl, a species that was considered to be at risk if the wholesale destruction of the Pacific Yew was allowed. Fortunately, human ingenuity came into play and the pressure was removed from the stocks of the Pacific Yew. An exploration of other *Taxus* species identified the needles of the yew (*Taxus baccata*), a widely grown ornamental shrub, as a source of a chemical structurally related to taxol. This chemical could be converted chemically into a close relative of taxol which was also an effective treatment for ovarian cancer. A huge programme of collecting the clippings of thousands of yew trees annually provided a viable source of the drug. Although synthetic routes to taxol have been reported, none has been successfully brought into commercial production, despite every considerable effort. Likewise, attempts to grow *Taxus* cells in culture have not yielded an alternative commercial source of the chemical.

The story of the discovery of taxol illustrates one of the major problems in seeking pharmaceutical agents among the hundreds of thousands of NPs made by plants. An effective, valuable chemical might be found but a practical, economic source of the chemical might not be. Indeed, a natural source of a very important drug could be very bad news for threatened habitats. How that problem might be resolved is discussed later.

Taxol holds another interesting lesson for us. As in the case of vinblastine or vincristine, it is unlikely that the Pacific Yew made taxol to gain fitness by making an anticancer chemical. It is arguable whether the fitness of plants is reduced by anything similar to cancer in animals. Some individual plants do have clumps of cells made by repeated division, commonly seen as galls, but these structures are usually the result of the invasion of the plant by an insect or a bacterium. There seems to be nothing analogous to the spreading of a cancer as found in animals and the author knows of no example of plant dying due to 'cancer'. Thus, there was no rational reason to seek anticancer drugs in the Pacific Yew, or indeed in any plant.

The annual total world market for anticancer drugs is currently about $50 billion and is expected to double within a decade.

The future of NPs as pharmaceutical products

The several examples given in the previous section of the extremely valuable NPs used as pharmaceutical agents are regularly used to justify more funding for NP research. Yet, despite the fact that NPs are still so important to the pharmaceutical industry, the collection of samples of NPs for screening for pharmaceutical activity declined over the past two decades.[12] What explains this lack of faith in NPs by the pharma industry? Have the past 20 years of drug discovery been but a temporary phase, when new ideas and new toys distracted attention away from the proven approach of seeking drugs in collections of NPs?

The loss of interest in NPs

There are several factors that have combined to make the screening of collections of NPs look a less attractive way of seeking new pharma products.

High-throughput screening

Large-scale screening of NPs is expensive and slow; it generates many false leads and it is intellectually dull. The only intellectual excitement that came to the subject in the 1980s and 1990s came from engineers who designed computer controlled robotic systems to dispense and analyse samples. These robots could not only do the boring work with great precision but they could also record and display the data. Biochemists, working with these engineers, devised biochemical procedures that could be miniaturised so that the enzyme activity in a sample, or the binding of a substance to a particular protein, could be measured in thousands of samples a day with little human effort. This approach became known as high-throughput screening (HTS).

Not only did the HTS robots operate 24 hours a day and 365 days of the year but the HTS approach also capitalised on the rapidly increasing knowledge of cell functioning. Drugs discovered in the middle of the twentieth century had nearly all been found using whole organism screens, with the eventual target of the drug being unknown at the time of discovery. However, as the mode of action of existing drugs was discovered, it was clear that the selective toxicity so essential for use was always based on some fundamental protein–ligand interaction (see Chapter 5) that could be analysed and understood. Could not the traditional approach be reversed? Instead of finding a useful biological action and then understanding the basis for that action, why not use the current knowledge of cell functioning to predict how to find agents that acted on the target process alone? The newly developed methodologies of HTS were ideal for such an approach. HTS depends on seeking a chemical that shows a particular kind of biomolecular activity. This seemed a very great advantage because it optimised the chances of finding the highly selective action that is the dream of all those seeking a new drug. For example, a screen of a collection of chemicals to find an anticancer drug that uses a cell multiplication assay will identify many chemicals that are toxic to some process in the cell and will thus stop the cell dividing—there will be many false positives due to the fact that the cell has many targets for chemicals that act specifically or non-specifically in a toxic manner. However, a screen that is based on microtubule functioning, or better still a specific aspect of that functioning, will find only chemicals that hit that target. The refined assay will find many fewer false positives.

All the major pharma companies invested heavily in HTS in the 1980s and 1990s, but the success of the engineers soon produced a new bottleneck. The capacity of the HTS instruments grew faster than the rate at which new chemicals were being made. The capacity to test samples for the very specific activity grew from hundreds of samples a day to tens of thousands of samples per day. It was soon very apparent that the more specific the target selected for study the lower the frequency of finding any significant activity when the chemicals were tested at low concentration (for the reasons explained in Chapter 5). Given that the cost of conducting every test was now very low, the new limiting factor in drug discovery was the size of the collection of chemicals available to test. Even during the early years of HTS, a library of 10,000 chemicals could be screened within a few days. The challenge switched from how to screen chemicals quickly, cheaply and efficiently to how to increase the size of the library of chemicals that a company possessed.

Combinatorial chemistry

For 150 years chemists had been trained to synthesise and purify new chemicals. The purity of the final product was an indication of the chemist's skill. The chemical agents used to bring about transformations in a synthetic sequence had been developed over the years to be good at producing high yields of the desired product. So most organic chemists working for pharmaceutical companies at the time when HTS procedures were developed were using their skill to make particular structures with a high purity of the final product. The structures being made were 'designed' to have properties that experience suggested were appropriate for high biological activity and these chemicals were delivered to those conducting the screen in a purified form. The testing of the activity of the specific chemical was the aim; hence, impurities would just confuse matters. However, there was already a mass of evidence available to show that it was very hard to predict which chemical structure would possess a certain type of biological activity. Experience suggested that after a 'lead' had been found (a lead is a chemical that possesses some activity of the desired type) knowledge could be used to make analogues of the 'lead' compound, analogues that might be expected to be even better than the original lead. Thus, knowledge was often more useful at optimising an outcome from an initial discovery than it was in finding the original lead. A simple, but very radical, thought emerged from this logic. Why not make as many chemicals as possible, in impure mixtures, and test the mixtures to find the lead and then work backwards to find the active compound once any mixture had been shown to possess the desired, but very rare, biological activity? This approach was to become known as *combinatorial chemistry*. The aim was to devise synthetic methods that could produce chemical diversity rather than single pure substances.

Molecular biology—another distraction from NP research?

Combinatorial chemists, HTS biochemists and engineers were not alone in taking funds away from those who had for decades being slowly gathering plant samples and painstakingly extracting and purifying novel NPs from them for screening. By the 1980s, genes were something to be found in vials, not just in organisms. The techniques and knowledge of the molecular biologists were rapidly assimilated by the pharma companies. Genes code for proteins and proteins were now something that could be made in fermentation vats at a reasonable price. Insulin and human growth hormone, very valuable substances for patients who cannot make sufficient themselves, were traditionally laboriously extracted from animal organs. By expressing human genes in microbes and using the fermentation methodologies (well known to most pharma companies because of decades of growing microbes for antibiotic production) to grow the genetically transformed microbes, human insulin and growth hormone could be manufactured for the first time. In theory, a company could realistically expect to be able to make any protein of clinical use, either to supplement deficiencies or to be used as a diagnostic tool. Furthermore, because patents could be taken out on novel genes, the discovery of the role of a gene in any form of disease or ailment could be a very valuable asset to a company. Seeking genes associated with appropriate diseases or ailments not only provided potential valuable diagnostic tools, but also opened the possibility of using the expressed proteins in a new HTS methodology to target those proteins.

Bioprospecting—new term for an old approach that attracts new advocates

At a time when the pharma companies were, one by one, reducing their commitment to the search for biologically active NPs in plant and microbial extracts, two quite different groups were advocating the opposite.

Some environmentalists, frustrated by the lack of public concern over the continuing destruction of many important ecosystems, realised that they might have greater success in preserving such ecosystems if they appealed to the self-interest of the public. Clearly, the public valued pharmaceutical products, many of which were NPs. In particular, the most dreaded human illness in the most affluent countries, cancer, seemed to be especially susceptible to treatment with NPs (taxol, vinblastine and vincristine). Put simply, maybe the next important anticancer drug would be found in some obscure plant living in some threatened ecosystem? Hence, there were both economic and humanitarian reasons to preserve such threatened ecosystems because these ecosystems have a high probability of containing organisms that will be able to produce the next generation of NPs needed by humans for drug use. The term bioprospecting was introduced to describe what had traditionally been called NP screening and it is a term that has become widely used by its advocates. The enthusiasm for bioprospecting can be judged by the fact that as of late 2008, 141,000 sites are found when searching Google for this word.

The general public readily picked up these ideas in a simpler form from the popular press. The idea that the next generation of miracle cancer cures awaited discovery in the rain forest, in coral reefs or in the deep ocean, certainly made many people take a greater interest in the preservation of these habitats. Furthermore, the arguments in favour of preserving biodiversity in order to retain the value in the chemical diversity gained support from some analyses by academic economists. Balick and Mendelsohn[13] studying the harvesting of medicinal plants from a rain forest estimated that annual revenues of $16–61 per ha could be achieved; hence, a high value could be placed on the rain forest for that use alone. Pearce and Puroshothamon[7,14] took that analysis further when they estimated that OECD countries might suffer an annual loss of £25 billion if the 60,000 threatened species were actually lost as a medicinal resource. The environmentalists, and those economists who had calculated the value of these as yet undiscovered chemical resources, were especially encouraged by the fact that there were some examples of bioprospecting in action. Two examples were quoted regularly as evidence of the value of bioprospecting. First, the widely publicised agreement by Merck and Co. to enter into a bioprospecting agreement with the National Institute for Biodiversity (INBio) in Costa Rica in 1991.[15] Second, the investment by Eli Lilly in Shaman Pharmaceuticals, a small company that aimed to use local ethnobotanical knowledge to target plants with a high chance of containing a physiologically active NPs. By the 1990s, it seemed that the pharmaceutical companies were not alone in renewing their interest in screening chemicals from the natural world. The US National Cancer Institute (NCI) had restarted its programme to look at NPs, despite the fact that the previous NCI bioprospecting programme (1955–80) had screened 200,000 extracts for anticancer activity with such limited success that the programme was run down. By 1995, the NCI had produced 40,000 extracts for screening, and out of that 18,000 extracts had been screened for anticancer activity. By that time about 1% showed some positive activity.

This apparent renewed interest in plant products as a source of pharmaceutical leads in the 1990s led optimists in the development community to identify an opportunity to build a revenue stream between the rich, health-conscious, but resource-poor (in biodiversity terms) nations and the poor, resource-rich less developed world. Discussions about bioprospecting moved on to consider issues of equity—how could the poor, developing nations negotiate a good deal with the powerful drug multinationals? How could any income stream that was negotiated be targeted at the most appropriate groups within the developing country (and who were such groups?). Much has been written about these equity issues[16] but less has been written about the logic behind the basic premise that bioprospecting is the best way of discovering drugs. Are rain forests, coral reefs or pristine oceans really a wonderful source of chemical diversity? More importantly, is this chemical diversity likely to contain the next generation of blockbuster drugs?

Among economists, there is still a debate regarding the rewards that can be expected from bioprospecting. Rausser and Small,[17] after a thorough theoretical analysis, concluded that using the accumulating ecological, ethnobotanical and biological knowledge it should be possible to make the screening of NPs much more rational and hence much more productive.

Bioprospecting—the reality

Drug development is more than drug discovery

Rausser and Small overlooked several factors in their economic analysis of bioprospecting and hence overestimated its potential.[18] Four of those factors are crucial and evidence for the importance of those factors was available by looking at the experiences of the big pharmaceutical companies.

The first problem with the analysis of Rausser and Small was that they overemphasised the cost of lead discovery relative to the total cost of bringing a drug to market (now estimated at several $ billions). As discussed earlier, the cost of screening samples has dropped dramatically, with less effort being needed to screen large libraries and improved screening methodologies having reduced the number of false positives being found. The major costs of drug development are now safety testing, preclinical trials and clinical trials. The industry has been seriously alarmed by the fact that very extensive safety testing still does not eliminate the possibility that a drug will reach the mass market before rare adverse side effects start to be recorded when the range and number of patients massively exceeds the number and range that can ever be studied in clinical trials.[19]

Second, Rausser and Small, like many who admire the ethnobotanical knowledge of herbalists, overemphasised the importance of ecological and ethnobotanical knowledge in facilitating the selection of the plants to collect. Although many undeveloped societies possess a very rich cultural knowledge about the use of plants and fungi for medicinal uses, much of that knowledge will relate to diseases or ailments that can already be treated in western society by existing drug treatments, hence there may not be a commercial need to find further treatments. Likewise, many conditions that are a serious concern to western societies, and for which there might be a very great need for a improved drug, might be conditions that the simpler society has never experienced, hence there might be no appropriate ethnobotanical knowledge. The diseases of the rich, overfed, possibly stressed urban westerners are not the diseases of the poor, rural, forest dwellers. The diseases associated with old age are, likewise, likely to be of less interest in communities where most individuals die in childhood. The most striking example of a mismatch between traditional shaman knowledge and the needs of a modern society is HIV/AIDS where a novel emerging disease must be tackled without relying on past experience. These arguments are not supposed to belittle or undervalue local ethnobotanical knowledge in any way; the arguments are made simply to indicate that some knowledge cannot be expected to guide the large pharma companies that target very different populations.

Third, and most crucially, Rausser and Small failed to appreciate that an active lead is often only useful if a practical, economically appropriate source of that chemical, or a biologically active analogue, is available. This is where synthetic chemicals usually have an advantage over NPs in any drug discovery programme. Self-evidently, a synthetic chemical made for screening purposes must be a chemical that could be manufactured. Although it might be difficult to make chemicals speculatively for screening purposes, the majority of those made and tested are likely to be structures that the chemist was confident could be made with reasonable effort on their part. Such chemicals are likely to be ones that can be synthesised if needed on an industrial scale at an affordable price. This line of argument is not an absolute one but one that relies on a balance of probabilities. Chemicals that are extremely hard to make are likely to be quite rare in libraries of synthetic chemicals. In contrast, when a mixture of NPs is tested in a screening trial, should a potent activity be found, it is quite likely that the isolation, purification and identification of the active principle will be hard and attempts to synthesise the compound, at a cost that can be borne by the market, might never succeed. Would penicillin ever have left the laboratory if it had been made by a microbe that was extremely hard to grow in culture? The examples of taxol or vinblastine also serve to remind us that chemists lag far behind plants in terms of their ability to elaborate some complex chemicals. Thus, a company feeding its HTS with synthetic chemicals can start with a more optimistic and realistic appraisal of the chances of actually being able to bring a product to market than a company feedings its HTS with extracts of NPs.

Fourth, and finally, Rausser and Small seem to have assumed, like so many scientists until recently, that organisms have evolved only to retain biologically active NPs, as if organisms were doing the first stage of a screening trial on behalf of humans. As explained in Chapter 5, the Screening Hypothesis, based on well-established physicochemical principles, postulates that most NPs are simply members of the NP library that the natural world has made. Like individual chemicals in the libraries of synthetic chemicals made by humans, most of the chemicals will possess no potent biomolecular activity.

To summarise the arguments, the majority of NPs found in plants and microbes are unlikely to possess potent biological activity and even less likely to contain specific, potent biological activity that could be usefully exploited for pharmaceutical use. Furthermore, even when a naturally derived chemical is found to give a good lead, the chemical complexity so characteristic of NPs may make commercial production expensive or impossible. It is surely significant that culturable microbes have been so important as producers of NPs of pharmaceutical value to humans because they can be selected to overproduce complex molecules that humans would find impossible to make.

Bioprospecting—the future?

Although the earlier discussion explains why in recent years the pharma industry have moved away from NP screening, that does not mean that NPs do not hold great promise as pharmaceutical agents in future. Indeed, the opposite is true. There is a growing acceptance that, as the industry enters the twenty-first century, the expectations of the HTS and combinatorial chemistry era have not been fulfilled. Indeed, the screening of hundreds of thousands of chemicals in the 1990s, chemicals made by conventional chemistry and by combinatorial methods using screens for many different targets, has produced an unimpressive list of major new drugs. Quite how bad the problem is cannot be fully quantified; telling the world that you have tested hundreds of thousands of chemicals and have found nothing of value tends to depress your share price and reduce the CEO's stock option value. However, there is now talk of the need to think again about ways to bring NPs into the screening programmes. The judgement being made is that all synthetic chemicals are all too often very similar to one another. The chemical diversity as drawn in two dimensions on paper, or indeed as represented in three dimensions using computer graphics, does not adequately convey the limited range of 'pharmacophore space' that is being accessed by the synthetic structures that are easily made by humans. As explained previously (Chapter 4), humans are good at building up simple carbon skeletons and quite good at building up more complex ones if they put enough effort into the task. Humans are also fairly able to elaborate these skeletons in a limited number of ways. But plants and microbes, using enzymes, can elaborate a much wider range of structures, and make more subtle and delicate elaborations, to produce a bewildering range of complex shapes in a huge range of sizes. Maybe, the 3-D complexity of some NPs is just what is needed to form a stable and high-affinity binding to a particular protein? So, how can one introduce NP diversity back into the screening programmes, without all the negatives discussed previously? Some argue that we simply need to turn the clock back and just put more effort into collecting NP samples and testing them. However, others see a more promising avenue to explore, combining ideas that have already been discussed.

Combinatorial biochemistry

The low frequency of finding biologically active molecules was the incentive to develop both HTS and combinatorial chemistry. As explained in Chapter 5, the goal of generating chemical diversity, using enzymes capable of acting on more than one substrate and possibly producing more than one product, is exactly how plants and microbes have evolved to optimise the generation of chemical diversity. Organisms making NPs make use of *combinatorial biochemistry*[20] to generate and retain chemical diversity. However, humans cannot access more than a small fraction of that NP diversity because most organisms making NPs occur in limited numbers, many such organisms cannot be cultured and it is hard for humans to find the one NP they could use among the huge numbers and amounts of valueless NPs that humans encounter when they extract an organism. So, how could this bleak situation be changed?

If the Screening Hypothesis (Chapter 5) is valid, it should be possible to enhance the generation of NP chemical diversity of an organism by adding to that organism a gene coding for another NP-making enzymic activity from a different organism (see Chapter 10). It matters not from whence that gene comes, it could be from a plant or microbe or even one of the non-specific enzymes involved in transformations of substances in the human liver for example. Such a genetic manipulation of NP-producing organisms to generate new NP diversity has already been reported (see Chapter 10). Consequently, bioprospecting might be carried out on laboratory generated organisms. This laboratory-based bioprospecting will have added advantages. First, it will be possible to use organisms that can be grown easily in the laboratory, organisms that can also be grown in large-scale fermentations. This will immediately address the problem of making any useful chemical once it has been found by screening because the need to make the chemical will have been taken into account at the first stage of the process. Second, this approach opens up a huge untapped potential to investigate and possibly exploit the biochemical potential in microorganisms that currently cannot even be grown in the laboratory, let alone in commercial production.[21] It is considered that only a very small fraction (<10%) of the microbes that exist in soil have ever been grown in isolation and this is hardly through lack of effort. The requirements for each organism are unknown and are likely to remain so. However, the DNA of such organisms is now accessible; hence, their biochemical capacity can be explored by incorporating parts of their DNA into other organisms that can be cultured. These concepts are already being explored and methodologies have been devised not only to engineer such genetic supplementation but also to screen the biological activity of any new chemicals that are made in an efficient miniaturised HTS process. This is bioprospecting in a new guise and one that is based on a random choice of genes rather than a random screen of chemicals. The genes that give the useful product are just as likely to be found in an insignificant microbe that will never be identified, let alone grown in culture, as in the most beautiful tropical tree. Furthermore, as our knowledge of the way in which DNA sequences influence protein structure and function increases, it will be possible to engineer the biosynthetic ability of organisms to change their spectrum of biosynthetic capabilities, hence the products they can make.

Drug metabolism and NP metabolism

Many animal species, especially herbivores or omnivores, evolved to cope with the presence of chemicals in their diet (see Chapter 6). Although specialist feeders, those that feed exclusively on some food source containing high concentrations of a few NPs, may have evolved to detoxify or become resistant to those substances, generalist seem more likely to use generic mechanisms which simply attempt to keep all NPs below a toxic level. The use of generic mechanisms, which are not selected on the basis of being optimised to reduce the concentration of one specific chemical but which are effective against a wide range of substances, means that when humans take pharmaceutical drugs, or administer veterinary drugs to their domesticated animals, these generic mechanisms will have a significant chance of acting on the administered substance. Some authorities now classify this form of metabolism as *xenobiotic metabolism*, meaning the metabolism of substances made outwith the organism carrying out the metabolism. However, in such discussions NPs are rarely discussed as a very significant evolutionary driving force that has shaped these metabolic properties.

It is also common in discussions of *xenobiotic metabolism* to find that there is an assumption that all xenobiotics must be toxic and that any metabolism of the xenobiotic must reduce the toxic load. Of course the reality is that most xenobiotics (other than drugs) will possess no significant biomolecular activity (for reasons discussed in Chapter 5) and the degradation of a substance creates one or more new compounds that are equally likely to possess significant biomolecular activity; hence 'degradation' is no guarantee of detoxification.[22]

As predicted in Chapter 6, from the perspective of NP evolution, those studying xenobiotic metabolism report generic mechanisms and they have classified the typical degradative sequence as being in three phases. Phase I, typically changes a substance into something more polar (more water soluble), commonly by the action of one of the common, non-specific P_{450} enzymes found in many organisms.[23] Phase II involves the 'conjugation' (the joining together of two entire entities) of the product of the Phase I reaction with a common substance such as glutathione or glycine, again making a more polar substance. Finally in Phase III, the conjugate is further modified or transported from the cell by specific proteins.

It may be time to refocus the discussions of the metabolism of xenobiotics to place the xenobiotic metabolic properties into the evolutionary perspective outlined in Chapter 9. That would place NP metabolism more clearly into the arena and would possibly help integrate discussions of drug metabolism into the broader biological context.

What does this chapter tell us about the way science works?

The role that serendipity plays in science is often a surprise to non-scientists who sometimes think that science is simply about the power of 'the scientific method'. Fleming's chance observation of the mould growing on his contaminated plates (penicillin); the chance that a brother sends a few leaves to a scientist seeking an anticancer drug (vinblastine); the chance that a technician finds the best antibiotic producer in a rotting fruit in a local market instead of in thousands of specifically sought soil samples (penicillin); and the multiple chances of individuals with two complementary ideas coming together to solve a problem. It is sometimes said that the most successful scientists are those who observe and exploit the unintended, chance events, rather than focusing on a route already planned with their own thoughts.

The personality of scientists, especially when fame or money enters their lives, suggests that scientist are humans first and scientists second. The fact that so many people involved in antibiotic discovery were treated so badly, or treated others so poorly, shows that Nobel Prize winners are not always noble.

Finally, the bioprospecting saga shows how bandwagons attract passengers with all sorts of motives and these bandwagons take a lot of stopping. Once a bandwagon starts rolling, a few respectable researchers sometimes hitch a ride simply because they need money to advance their work and careers.

8

The Chemical Interactions between Organisms

The hypotheses we accept ought to explain phenomena which we have observed. But they ought to do more than this: our hypotheses ought to foretell phenomena which have not yet been observed.

—*William Whewell (1794–1866) English mathematician, philosopher*

Summary

NPs play a very important role in determining the interactions between individuals (of the same or of different species) that cohabit an area. The interactions between plants and animals provide examples of the way in which NPs play a role in interspecies interactions. In many animals, the key senses of taste and smell have evolved to be acute sensors of a very few NPs but most NPs are quite possibly never sensed by any organism. However, given that NPs evolved billions of years ago, and terrestrial animals and plants only about 400 million years ago, there is a very large hole in our understanding of the selection forces in microbes that drove the evolution of NPs for the majority of evolutionary time.

NPs and animal behaviour

Animal behaviour fascinates humans; it is a rare week when one cannot watch TV programmes illustrating the many weird and wonderful ways in which organisms interact. This fascination with animal behaviour stretches back throughout our history. Human knowledge about the interactions of organisms was, and still is, valuable. Every young child is not only encouraged to watch examples of animal behaviour but they are also told stories about animal interactions, encouraged to mimic animal behaviour and given toy animals. This is not surprising because it is hard to think of many human activities where some knowledge of animal behaviour does not benefit humans; farming, hunting, navigating, fishing and gardening are obvious examples; warfare, building, banking are less obvious ones. Consequently, of all the sciences, biology is the one that has the most immediate connection with people because everyone has considerable experience of animals, and to a lesser extent plants, from childhood.

The professional study of the interactions between organisms began in the nineteenth century but blossomed in the twentieth century. Not surprisingly, the way in which the mammals interacted initially dominated the subject. The emerging, very popular, subject of *animal behaviour* is largely focused on a few large animal species that fascinated humans. Simple descriptions of animal behaviour in their natural environment were followed by laboratory studies aimed at understanding both the mechanisms used by animals to sense information and the ways in which such information was processed and acted upon by the brain. Of the non-visual senses, smell and taste were soon understood to be important in providing information which initiated specific patterns of behaviour. Our own human experience tells us that taste and smell can provide very important clues about what to eat or what not to eat. By the mid-twentieth century, it was clear that NPs played an important role in the interaction between organisms, a role that was maybe underappreciated because, although humans have particularly impressive senses of vision and hearing,[1] they possess a much less impressive sense of smell. A huge scientific literature now exists which describes the interactions that occurs between tens of thousands of organisms. The diversity, complexity and subtlety of individual interactions cannot be fully summarised in this chapter, rather an attempt will be made to identify a few common principles that underlie such interactions. The main theme of the chapter is that NPs have been central to the evolution and co-evolution of many species.

The role of NPs in governing the interactions between organisms

Living creatures use, metabolically, a very wide range of chemicals but NPs are not significant in this respect. The reason why NPs were once considered secondary (Chapters 1, 5 and 9) was because an individual in a population can survive in the short term when it is not making, or accessing, NPs. However, it has been established that while individuals can survive without NPs, individuals that have evolved to make or sense NPs are fitter. Why might that be?

NPs—one fundamental action, many outcomes

At a molecular level, those NPs that possess potent, specific biological activity act in the same way; each NP associates with its own target protein. Because each target protein is embedded in an organism whose functioning depends on the 'correct' functioning of thousands of proteins at any one time, the outcome of the interaction of any one NP with any one target protein can be expected to be characteristic but hard to categorise. An NP interacting with a protein that is an enzyme might be expected in the majority of cases to have a negative effect on the capacity of that enzyme, hence reducing the rate of synthesis of whatever that enzyme makes. An NP interacting with a protein involved in a sensing might increase the output of the sensing cell. However, to further complicate the outcome of such interactions, the inhibition of an enzyme need not always make an organism less fit and the stimulation of a sensor need not always make the organism act positively. For example, some pharmaceutical drugs that enhance the fitness of individuals act by inhibiting specific enzymes. Likewise, some organisms react to sensing a smell by moving away from the source of the smell or rejecting the food that has a particular taste—is that a positive or a negative response?

From these considerations, it is predictable that every organism with a capacity to make NPs will possess, in theory, the potential to influence the metabolism or behaviour of every other organism that shares its habitat. Those other organisms will possess between them thousands of proteins that are potential targets for NP action (Figure 6.3). However, there will be many constraints that limit the evolution of NP-mediated links between the NP producer and other species.

- Many of those proteins in a habitat will be ones that are highly conserved between species; hence, the organism making an NP will possess a significant number of target proteins itself—the potential for 'autotoxicity' will be considerable.
- Because many proteins will have been significantly conserved within certain groups of organism, selective action between individual species within that group may be difficult to achieve. For example, an individual plant that, by mutation, produces a novel NP that has powerful insecticidal properties will not automatically gain fitness because insects visiting the plant will not necessarily be harming the plant; many plants rely on insects for pollination; many plants gain fitness by attracting wasps that parasitise the eggs of insect herbivores. Consequently, the number of potential target proteins available to the plant is constrained by the fact that large groups of organisms are sufficiently alike that selective action against individual species is hard to achieve. Indeed, once any NP-producing organism starts to develop positive relationships with any other group of organism, they may be making it harder to use NPs to target any similar organisms that have a negative effect on them.
- Most of the organisms that share a habitat with other species interact specifically with a very limited range of those species. Individuals of a particular plant species might grow adjacent to hundreds of other plant species, but the only interaction between one plant and its neighbours might be generic competitive strategies, strategies that would be used when growing in competition with any species, including its own. Likewise, an individual plant may be visited by many different insect species but most insects will simply visit en route for some other destination and have no meaningful interaction with the majority of plant species they encounter because many insect species are specialists that have no interest in any organism other than the few they have evolved to interact with. A cabbage white butterfly flitting around the garden is not enjoying the scents or flavours of all the plants it encounters; it is simply seeking brassica plants in what must seem to it to be a rather frustrating world. One can conclude that an individual making a novel NP will only gain fitness if by doing so it can influence the fitness of a very few key species with which it really interacts—those species that significantly influence its own fitness.

In summary, every organism making NPs has a much smaller number of potential target proteins available to it than one might initially expect. However, it is inevitable that, just by chance, some organisms will make the occasional NP that has a potent ability to bind to some protein in an organism with which it never interacts. Such fortuitous interactions will be selectively unimportant to the producer of the NPs as they do not add or detract directly from the fitness of the producer.[2]

With this short general introduction to the chapter behind us, it is time to look more closely at some specific interactions that involve NPs. Given that NPs evolved in microbes, it is necessary to start our analysis with those organisms, despite the fact that the interaction between microbes has been subject to much less attention than the interactions of higher organisms.

Microbial interactions—did NPs evolve to play a role in 'chemical warfare'?

The capacity to make NPs evolved in simple microbes (see Chapter 3). However, many single-celled organisms can survive, alone, without any interaction with any other living organism, so why did chemical interactions between organisms evolve? This is a question that, like most evolutionary questions, cannot be answered with certainty but it is a question that deserves some speculative thought.

The simple answer might be that *life* is not simply about an individual merely living. *Life* for an individual is often about competing for resources, with other individuals of your own species or individuals of other species. *Life* for a species is about individuals surviving, reproducing and passing on their genes to another generation. Consequently, an individual organism that can detect resources more efficiently, that can detect hazards in their locality and that can find a 'mate' efficiently will be fitter than individuals possessing lesser capacities in these respects.[3] For a microbe, the advantages are clearly going to be subtle. Very early in evolution, it seems possible that an individual microbe might have gained fitness simply by releasing a chemical that inhibited the fitness of a nearby competing individual. However, this simple idea needs to be pondered a little because it may be too simplistic as stated. It is too easy to think of microbe 1 producing 'an antibiotic' to kill adjacent microbe 2; hence, microbe 1 now has access to the resources it previously shared with microbe 2. Yet when microbial species have been screened by humans for NPs with antibiotic properties it has been extremely hard to find such NPs (see Chapter 7). But this very low success rate in antibiotic discovery should not allow us to reject the model that some microbial species might have evolved to gain fitness by producing NPs with antibiotic properties. It could well be that the NP properties needed to make a good antibiotic for human use are simply not equivalent to the NP properties needed to allow a microbe, in some odd niche, to gain fitness from inhibiting another microbe sharing that niche. The ideal antibiotic for human clinical use is a chemically and metabolically stable substance that has a broad spectrum of activity and has no adverse effects on humans. Each or all of these properties might be unimportant to some odd microbe. So may be an effective antibiotic for a microbe is not equivalent to an excellent antibiotic for medical use and our searches for good antibiotics for human use simply miss the good antibiotics for microbial use? Let us explore this idea because there are some concepts that need to be clarified so they can be used later in the chapter:

- Gaining fitness is not simply a matter of killing enemies. Human experience tells us this. 'Economic warfare' is far more common in human society than armed warfare. Within an apparently stable human society, some individuals will be more successful at passing on their genes than others, yet rarely will this involve one person killing a competitor. Gaining more than your fair share of resources, and gaining access to the best genes available in potential mates (usually helped by having more than your fair share of resources), is good enough to help your fitness. Thus, a microbe can gain fitness by inhibiting a competitor's growth rather than killing the competitor outright. In a prolonged competition in the soil, a gene that gives a mutant an advantage of a few per cent could become widespread after a few generations. However, the chemical being produced as a result of possessing that gene might not seem impressive when tested for its ability to kill microbes pathogenic in humans.
- The competitive regime used in a pharmaceutical company's antibiotic trial might not provide a good model of the competition that occurs in the natural habitat of the microbes. For example, the rich nutrient media on which the potentially pathogenic microbes are grown in the laboratory are highly unlikely to mimic the nutrient conditions found in the soil, even if the laboratory conditions would be more equivalent to the rich supply of nutrients available in the human body. Thus, the outcome of competitions between individuals will be highly dependent on the circumstances under which the race is conducted—human experience shows that a race run on the athletic track might produce a winner who is poorly equipped to win a race up a rough, wet hillside.

These speculations provide some possible reasons why the low frequency of occurrence of 'antibiotics' in microbes cannot be used as a conclusive argument that microbes must have evolved NPs to serve a role other than competing with each other. Could it be that the term 'chemical arms race' often used when discussing NP evolution is really less appropriate than 'economic arms race'?

Although the only concept most people have of microbial colonies comes from photographs of colonies growing on the flat surface of a nutrient agar plant, many microbes in the natural environment live in mixed communities, sometimes in 'biofilms' (Figure 8.1).[4] The microbial community produces its own structured environment which can sometimes protect some members of the community from human attempts to eradicate the microbes (cleaning water pipes is bedevilled by this problem). It seems possible that biofilm properties might have evolved to protect the individuals from predation by the many organisms that feed on microbes. It is possible that some rudimentary signalling takes place between individual bacteria, of the same

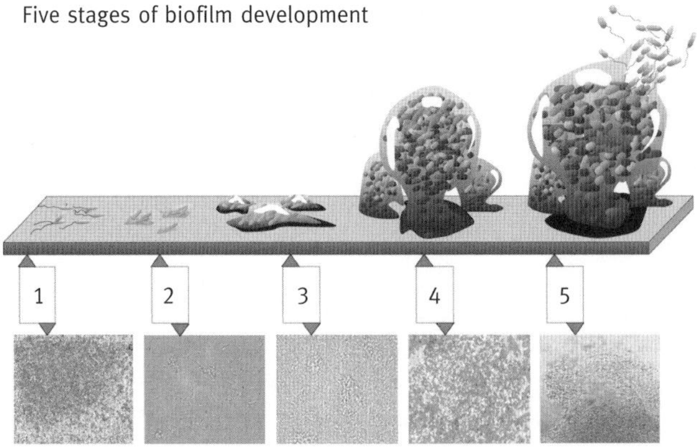

Figure 8.1. Biofilms are microbial communities that are made up of several species and often possess a distinct spatial structure. It has been found that living in a biofilm can offer significant protection to individuals, with sterilisation agents and even antibiotics being less potent against individuals if they are found in a biofilm. This is significant because it has been estimated that nearly three quarters of bacterial infections involve microbes that live in biofilm communities. Although there is a growing literature on the effects of NPs on biofilms, there is currently little knowledge of the production and metabolism of NPs in biofilms.

or different species, in such biofilm communities and NPs could play a role in these circumstances.

So, it cannot be denied that our knowledge of the role of NPs in microbes is very poorly explored and relies heavily, may be much too heavily, on ideas that have been embedded in our minds when studying the role of NPs in higher organisms. It may be useful at this time to refer again to an idea that was introduced in Chapter 6 where it was noted that microbes must possess a capacity to degrade many NPs. As argued in that chapter, the annual world synthesis of NPs is such that unless there was an 'NP cycle' the total world photosynthetically available carbon would have been locked up in NPs within a few centuries. There is no evidence for a significant accumulation of carbon in intractable NPs and given that most plant biomass (containing a few per cent of NP) passes directly to the soil rather than being ingested by herbivores, microbial degradation of NPs must be the most significant part of the NP cycle.[5] Given the broad substrate tolerance of the enzymes that make NPs it is possible that some of these enzymes participate in the degradation of NPs. Such a capacity would provide a microbe with an ability to reduce the concentration of any potentially inhibitory NPs produced by other microbes. If a microbe lives in a location where the flow of water is minimal, there will be a very limited opportunity to cope with toxic substances by dilution (the basic mechanism common to organisms with an excretion system with a route to isolating the excreted material or excreting it into a large volume of water). This thought leads to the idea that maybe evolutionary arguments about the role of individual NPs in microbes are too narrow and inappropriately focused. It might be more productive to think about the advantages that the possession of what we might call 'NP metabolism' might bring to microbes. The fitness benefits that might accrue to a microbe from possessing the versatile 'NP metabolism', the ability to make and to degrade chemical diversity, might be hard to pin down if one focuses only on a very few NPs made by one species under a limited number of conditions. Maybe the NP metabolism in microbes is akin to the immune system in higher animals—it is the net benefit from the possession of a versatile capacity that is more important than the value of any single product produced by that capacity at any time.

Multicellular organisms making and responding to NPs

How might a multicellular organism gain fitness by producing an NP?

Once organisms that could move evolved, and 'behaviour' evolved,[6] the opportunity of certain organisms to gain fitness by making NPs may have increased considerably. Indeed there does seem to be a rough division into two groups of organisms—sessile organisms that are rich in NP metabolism (plants, fungi and bacteria) and motile organisms that possess little or no capacity to make NPs but have a capacity for rich and diverse behaviour (animals).

Why might the opportunity to exploit NPs increase after movement and behaviour evolved? The evolution of major new faculties in organisms inevitably results in new gene products being made. Each new protein will have a probability of possessing sites at which NPs can bind such that the ability of the new protein to perform its function will be impaired. Not only will there be an increase in the number of potential NP targets, but as evolution proceeds, it is likely that these new target proteins will be increasingly unlike proteins in the NP-producing organisms. Consequently, it is predictable that organisms with a nervous system will possess a number of proteins that are absent from most NP-producing organisms. As explained in Chapter 5, the organism making a new NP is the organism that is most susceptible, suffering a negative effect owing to the presence of that new chemical; the maker of the new NP will be exposed to the highest concentration of that substance. Two interacting organisms with very similar protein compositions will inevitably have a very low probability of producing a new NP that will reduce the fitness of the receiver more than it reduces the fitness of the maker. However, an NP producer that is interacting with an organism that has a significantly different protein composition will have more opportunities to gain fitness by targeting one of the proteins in its competitor that is unlike any of its own proteins. Consequently, it is predictable that NPs that interact with proteins evolved for specialised functions in the nervous system or used in sensors linked to the nervous system (e.g., have an effect on behaviour) might have been significantly favoured by evolution. Certainly, human experience would support this argument. For example, many of the major attractive aspects of NPs for human use (Chapter 2) are linked to behavioural rather than physiological effects of the NPs (Table 8.1). There is also other evidence that the nervous

Table 8.1. Many of the NPs of interest to humans (see Chapter 2) bind to proteins which play a part in the central nervous system.

Substance	Species	Major valued NPs	Receptor binding NP (action of NP)	Endogenous ligand
Tobacco	*Nicotiana* sp.	Nicotine	Nicotinic (agonist)	Acetylcholine
Coffee	*Coffea* sp.	Caffeine	Adenosine (antagonist)	Adenosine
Tea	*Camellia sinensis*	Caffeine, theophylline, theobromine	Adenosine (antagonist)	Adenosine
Chocolate	*Theobromine cacao*	Theobromine	Adenosine (antagonist)	Adenosine
Opium	*Papaver somniferum*	Codeine, morphine	Opioid (agonist)	Endorphins
Cannabis	*Cannabis sativa*	Δ9-THC	Cannabinoid (agonist)	Anandamide
Coca	*Erythroxylum* sp.	Cocaine	Dopamine	Dopamine
Khat	*Catha edulis*	Ephedrine, cathinone	Adrenergic	Norepinephrine, epinephrine
Betel nut	*Areca catechu*	Arecoline	Muscarinic	Acetylcholine

Source: RJ Sullivan et al. (2008). *Proc. Roy. Soc. B*, 275, 1231–41.

system is remarkably vulnerable to chemical impairment. For example, when humans decided to seek insecticides, the screening methods chosen seemed to have a heavy bias towards finding nerve poisons. Each of the three major generations of insecticides used in the twentieth century acted on the insect nervous systems.[7]

Another threat—muscle power

The evolution of movement and behaviour also opened up a huge range of threats and opportunities to the land plants. It is not practical to consider all the range of species interactions in this chapter; therefore, examples will be drawn from the interaction of plants and insects.

Movements of parts of the organism (muscles in abdomen, in legs, in wings, in the jaw) gave herbivores new capacities to exploit plant material of all types. For example, individual insect species evolved to physically enter and move within every type of plant organ (seeds, leaves, stems, roots, flowers) and the jaws enabled the insects to physically disrupt tissues and cells to gain access to the nutrients inside the cell (and enabled microbial symbionts in the insect gut to access the ingested cell walls). The ability of the insect (or its parent) to move between plants enabled populations to spread rapidly. Although an individual insect rarely consumes a significant amount of plant material, because insect populations can increase very rapidly, insects sometimes present as great a threat as an individual large mammalian herbivore. An individual cabbage white caterpillar can eat only a small per cent of a cabbage leaf but a population of caterpillars on a single cabbage plant can consume a significant amount of leaf biomass over a longer period of time than a passing large mammalian herbivore. If the insect herbivore attacks the young growing leaves or the apex of the plant, it is not the loss of current plant matter that is significant but the loss of the future photosynthetic area. Likewise, insects that lay their eggs in young developing fruit are destroying future potential by consuming very small amounts of current production.

The evolution of NPs to counter new threats

At each stage of the interaction between an individual plant species and an individual insect herbivore species, there are opportunities for the plant to gain fitness by influencing the behaviour of the insect as well as directly reducing insect fitness by attacking the basic physiology of the insect. But before considering this further, one has to address the concept of cost–benefit analysis.

Costs and benefits of defence

Making and maintaining a defensive system clearly involves a cost—that applies to an individual human, to human communities, to nations and it applies to all organisms that have evolved any form of defence. In human societies, it is easy to identify the troops, tanks, warplanes and warships but even in human societies it is hard to find unambiguous evidence that the existence of these resources produce the benefits claimed. In human societies, where the cost of defence can be calculated with some accuracy, the 'opportunity costs' (the cost to society of not using that resources devoted to defences in some other way) are very hard to evaluate and the 'deterrence benefit' (the benefit that supposedly accrues from the possession of the offensive and defensive infrastructure) almost impossible to evaluate. Humans studying the cost and benefits of defence systems in other organisms have also found it much easier to identify various components of the defence systems (e.g., thorns, hard structures, hairy leaves, bitter taste, etc. in plants) than they have in evaluating the cost of producing and maintaining defensive chemicals (let alone the opportunity costs and the deterrence benefit).

The most complete analysis has been performed on higher plants, and an extensive literature exists.[8] Because individual plant species can have very different life cycles, it has been tempting to seek contrasting life cycles where it could be predicted that different cost–benefit ratios would be expected. For example, it was suggested that short-lived annual plants (ephemerals) would invest less in defences than long-lived perennial plants. The logic was that ephemeral usually come from a large seed bank[9] and each year, or at intervals during the growing season, many individuals can appear, grow rapidly and produce tens of thousands of seeds within weeks or months. Under such circumstances, a heavy investment in defence might not benefit an individual because the chances of being subjected to attack is low and the investment of resources in growing rapidly to set seed quickly might be more productive than making rarely used defences. In contrast, a tree has to exist for years or decades in the same place before it produces seed; consequently, the tree provides an annual opportunity for any adapted insect pest. However, on the other hand, the tree is running an account that compounds annually so it can afford to defend itself more fully.[10] It is true that for the obvious physical defences ephemeral plants do seem to lack the physical defences that are obvious in some perennials but there are some perennial herbs where this conclusion is much less secure. It is unlikely that there are simple universal answers about the cost–benefit balance, but it is universally agreed that producing and maintaining defences that are unused for long periods will make an organism (or nation) less fit. Consequently, it is predictable that any generic methodology that helps reduce the costs of defence will bring a benefit to those that use such generic mechanisms. One such mechanism is 'inducibility'.

Inducibility of NPs

History tells us that human societies will massively increase defence spending when they are being attacked or feel that they are about to be attacked. What might seem to be ruinous levels of military spending in peacetime seems worthwhile when war is imminent. In other words, humans have found that maintaining a dormant capacity to make weapons when needed is valuable in uncertain times. An analogous strategy seems to have been evolved by other organisms.

The fact that many plant species respond to fungal and/or insect attack by making more NPs has been taken as evidence that plants have evolved 'inducibility' as a way of reducing the cost of making NPs by making them in significant quantities only when needed. This is an appealing idea and this ability of organisms to vary their rates of NP synthesis has been a very important subject for study during the past 25 years. However, it has been difficult to provide conclusive evidence that the inducibility of NPs was evolved solely as a cost-reduction strategy. For instance, Agrawal and Karban[11] list a number of alternative hypotheses to explain the fact that some NPs increase in concentration in certain plants after insect or fungal attack. For example, one model for the evolution of inducibility is that the gains to the producer of NPs come from the lack of NPs when uninduced rather than the increased NP level after attack. This model is based on the fact that many insects find the plant species they seek by finding the source of NPs that is characteristic of that species. Consequently, a plant producing lower concentrations of NPs will be harder for an insect to find and it will be subject to lower rates of attack. Another model to explain the evolution of inducibility is that inducibility helps reduce the selective pressure that drives the evolution of resistance mechanisms in the insect (or fungus)—an analogy in human experience is that the development of resistance to antibiotics can be reduced by avoiding the exposure of bacterial populations to high levels of an antibiotic unless a serious threat to health exists. It seems very possible that the inducibility of NP synthesis was not evolved or maintained by a single selective pressure.

Inducible defences against fungal attack

The idea of 'acquired immunity' was widely accepted by animal biologists by the end of the nineteenth century, but the idea that plants might gain some immunity after challenge was much slower to gain hold.[12] However, in 1940, KO Müller and H Börger published evidence that potato plants, after infection by potato blight (*Phytophtora infestans*), seemed to have some acquired resistance to further infection. They speculated that the plants produced an antifungal substance after infection—a substance identified some years later as the sesquiterpene rishitin.[13] The term *phytoalexin* is now used to describe an inducible antifungal substance made by a plant. In the decades that followed, several more phytoalexins were found, one of the most notable being pisatin (a compound formed by the phenylpropanoid pathway in peas). Pisatin was isolated when it was shown, with elegant simplicity, that a drop of water recovered from the surface of an infected pea pod had a greater capacity to inhibit the growth of fungi growing on an agar surface than a drop recovered from a healthy pea pod.

By the last quarter of the twentieth century, the concept of phytoalexins was well established but the total number of different phytoalexins recorded was still very limited. Furthermore, the discovery of an antifungal NP which was made in greater amounts after fungal attack was sufficient to give the chemical the status as a phytoalexin, despite the fact that evidence was often lacking that the substance really did play a part in defending the plant from which they were extracted. Phytoalexin research tended to concentrate on finding out much more about the few phytoalexins known rather than broadening the search for new examples. Consequently, a great deal is now known about the biosynthesis of each of the well-known phytoalexins and the genes involved in the pathways have been identified and characterised. By the end of the twentieth century, the mechanism by which cells detected fungal attack, and respond to that attack by making specific phytoalexins, was known and the concept of *elicitors* was developed. *Elicitors* are chemicals that are detected by the plant cell that indicate that a fungus is attacking cells locally. These compounds have been found to be either fungal-derived or plant-derived (e.g., bits of degraded plant cell wall polysaccharide which indicated that some organism was hydrolysing the plant's cell wall as it tried to invade) but a full discussion of these interesting compounds is beyond the scope of this book.

Inducible defences against insect attack

It is may be not surprising that those working on plant defences against insect attack were less drawn to the idea of induced defences. A comparison of the way in which a fungus or an insect attacks a plant suggests very significant differences. Fungal infection of a plant usually starts with a single, very small fungal spore germinating on the surface of the plant and the resulting fungal mycelium needs to penetrate the plant before the fungus can get access to the nutrients it needs to grow. Thus, the initial interaction of the fungus and the plant involves just a few plant cells and takes place over several hours. In contrast, when a herbivorous insect encounters a plant, the insect contains sufficient energy to sustain it for many hours and it does not need to grow to have the capacity to cause considerable damage to the plant. A herbivorous insect, even one recently emerged from an egg, can start to attack the plant within seconds and tens of thousands of plant cells can be consumed by the insect in minutes, well before those cells can initiate any chemical changes to deter the insect. There would seem to be little time for an inducible defence to work against insects; consequently, it was not surprising that many of those studying plant–insect interactions accepted that plants would need to defend themselves against insects continuously. The fact that many plant tissues did seem to contain quite large amounts of some NPs (e.g., phenolics and lignins), even when healthy, supported this view. However, by the 1970s, the idea of inducible defences against insect attack began to develop.[14] The driving force behind this change in thinking was not a conceptual advance but simply advances being made in an unrelated discipline. Analytical chemists were developing generic methodologies that were much more rapid, sensitive and precise (gas liquid chromatography, liquid–liquid chromatography and mass spectroscopy). The driving force behind these massive improvements in analytical techniques was the human need to measure pesticides, pharmaceuticals and illegal drugs more reliably, sensitively, conveniently and cheaply. The large amounts of money flowing into analytical chemistry were directed at methods optimised to detect minute concentrations of specific chemicals in biological samples. Given that some pesticides, pharmaceuticals and illegal drugs were NPs, or closely related to NPs, the spin-off benefits to the academic studies of NPs were considerable. Soon reports of insect-induced changes in NP composition began to appear and the concept of inducible defences against insects was soon accepted. Indeed, there was soon a tendency to use inducibility as an indicator of function. If chemical X increased following insect attack, then X must be involved in defending the plant against insect attack. This was rather suspect logic and in contrast to the phytoalexin story, few attempts were made to specifically seek insecticidal activity in plants that only appeared after insect attack. However, the concept of inducible chemical defences against insect attack advanced significantly when it was shown that when an insect was feeding on the leaves of some species, the NP composition was changed even in the unattacked leaves. The insect-induced changes were not only local but were also *systemic* (meaning a signal of some type moved from one part of the plant to another). Furthermore, the changes in NP composition in the unattacked leaves were not due to NPs moving from the attacked leaves to the unattacked adjacent ones, but to the unattacked leaves making their own NPs in response to some signal that passed out of the attacked leaves into adjacent tissues. It was postulated that 'systemic signalling' could enable the plant to gain fitness by increasing the chemical defence in leaves adjacent to existing sites of insect attack. The leaves under attack were not being efficiently defended but the leaves that insect might move on to within hours could be made less attractive. Likewise, if an insect arrives on one leaf, the chances of another individual from the same insect population arriving on the same plant at a later time increases; hence, preparing other leaves to deter other individuals of the same attacking species might increase fitness. Hence, the concept of an inducible chemical defence against insect attack accepted that some short-term loss of tissue to an individual insect was tolerable if the longer-term chemical changes were sufficient to reduce the fitness of that individual insect or its relatives. However, how do these simple, widely accepted ideas about inducibility fit into the overall model of the evolution of NPs?

The Screening Hypothesis and inducibility

As explained in Chapter 1, the study of NPs became seriously fragmented. For example, those interested in the role of NPs in defending plants against insect attack gave rather little attention to the results of those studying the role of NPs in defending plants against fungal pathogens (neither group of researchers bothered much about the role of NPs in microbes). Thus, the concept of inducibility gained wide acceptance among those studying plant–fungal interactions long before it was given similar attention by those studying plant–insect interactions. However, after the fungal- and insect-induced inducibility of NPs had been shown, both groups of researchers began to ask question about how the plant sensed attack and soon some common issues were being addressed. When a fungus attacks a plant it must gain access to the interior of the plant; hence, at some stage of the invasion process some damage must occur to the plant cell walls. Likewise, when an insect attacks a plant, physical damage to the cell walls will be inevitable. Even an aphid skilfully inserting a minute flexible pipe (the stylet) through a stem to penetrate the phloem, to gain access to plant's nutrient rich sap, causes some physical damage. This recognition that physical damage was a good indicator of attack was made at a time when plant biologists had already discovered some other ways in which plants respond to physical damage. For example, it was known that physical damage to plant cells was often accompanied by a rapid rise in the production of the gas ethylene (ethene), a compound that was known to have profound effects on plant growth and development. Not surprisingly, studies that pathogen or insect attack increased ethylene production were soon reported. Because it was known that ethylene produced a number of biochemical changes in plants, the possibility that ethylene played a part in a generalised 'attack response' gained currency. However, studies of other biochemical changes in plants subject to attack yielded two other chemicals that frequently changed at some stage after the attack started. Jasmonic acid (JA) (a chemical first isolated and characterised as an endogenous plant growth regulator) was eventually recognised as playing an important role in the 'attack response'. However, things became even more complicated when it was shown that the concentrations of salicylic acid (SA) were frequently found to be changed in plants subject to insect attack and it became the third partner in the what became known as the attack–response cascade. The two fields were drawn more closely together at the end of the twentieth century when methodological improvements made it easier to compare the protein composition, and/or the messenger RNAs coding for those proteins, in healthy plants and those subjected to insect or fungal attack. It soon became apparent that when an insect attacks a plant, the plant responds by increasing the transcription of a great many genes and decreasing the transcription of many other genes. For example, in one study,[15] *Arabidopsis* plants attacked by phloem-feeding aphids (*Myzus persicae*) increased the expression of 832 genes and downregulated 1349 genes. However, similar large changes in transcription were also found when the plants were exposed to a pathogenic leaf bacterium (*Pseudomonas syringae* pv. *tomato*), a pathogenic leaf fungus (*Alternaria brassicicola*), tissue chewing caterpillars (*Pieris rapae*) or cell-content-feeding thrips (*Frankliniella occidentalis*) and there was a very large overlap (50% in some cases) between the changes induced by the very different organisms.

Why is there such an overlap between the defence responses to such very different organisms? Clearly, some of the overlap could simply be the result of the common element of physical damage and studies of the response of plants to physical damage which mimics the damage of insect supports this. However, it is clear that the plant also responds to specific inducers from the attacker (e.g., insect saliva or chitin in pathogen cell walls) and even those 'specific' responses also show some degree of overlap. This overlap has been described as 'crosstalk', using the analogy of the unintended interaction that can occur between two or more electrical signals. However, the borrowing of the term crosstalk, a term that emphasises the unintended consequence of an interaction, might be highly inappropriate. Surely, efficient defences would be specific ones, responding only to stimuli that indicate a particular threat—insect, fungi, bacteria or mammalian herbivore. Some of those who first found such 'crosstalk', by accident rather than as a result of a specific search for such flexibility, rationalised the interaction of the defensive systems in terms of the need of an attacked plant to prepare itself not only to the primary attack but also any subsequent opportunistic attack by another organism. Clearly, an insect chewing a leaf breaks through some of the defences that protect the plant from fungal invasion; hence, it is plausible that the simultaneous induction of anti-insect and antifungal defences would occur. However, this explanation is less convincing as an explanation of the 'crosstalk' that occurs after fungal attack because a plant invaded by a fungal pathogen might not be more susceptible to insect attack—the converse might be true. So, the 'multiple dangers' explanation offered to explain 'crosstalk' is a reasonable hypothesis but there is an alternative explanation offered by the Screening Hypothesis.

Crosstalk—it is predicted by the Screening Hypothesis

The evolutionary constraint that lies at the heart of the Screening Hypothesis (that any molecule has a low probability of possessing potent biomolecular activity—see Chapter 5 for detailed arguments) must have influenced the evolution of inducibility. The reasoning behind this statement requires a recap of some principles.

The Screening Hypothesis drew on the human experience of making synthetic chemicals and testing these chemicals for their biological activity (screening). The first lesson learned was that, if seeking a chemical with a specific, selective effect on one target organism, one would expect to screen thousands of different chemicals before finding one that had the desired effect when applied at low concentrations. The second lesson was that once an organism has evolved resistance to one chemical control agent, those resistance mechanisms have a high probability of being increasingly effective in protecting against chemically related control agents (group resistance). The third lesson was that if one synthesised a chemical as part of a programme seeking an insecticide, it was worthwhile testing the same chemical as a herbicide, fungicide or indeed for any pharmaceutical effect because the type of biomolecular activity that might be found was only weakly predictable. There are many examples of chemicals made as part of a specific search for one type of biomolecular activity that were valueless in their hoped for original role but turned out to be very valuable in a quite different role.[16]

These three lessons have implications for our thinking about the evolution of inducible defences in organisms. Consider an individual plant species at any moment in evolutionary time. A mutant arises with a slightly changed NP composition. If the mutated gene is to be retained in the population, one of the new NPs being made must give a fitness benefit at a bearable cost. But the cost–benefit ratio could be very much influenced by the previously evolved ability to control the costs by inducing the production of NPs only when the maximum benefit can be achieved. In other words, at some stage in evolution, inducibility becomes a significant inherent factor in shaping NP pathways. If the production of a new NP cannot be enhanced after insect attack, then that NP has to have a much, much higher biomolecular activity than an inducible NP that can be made at higher concentrations only when needed. Thus, a new NP that is linked to an existing inducible chain will have a higher (but still very low) probability of passing the cost–benefit test than an NP made at a constant rate because it is not linked to the previously evolved induction processes (if the rate of synthesis is low, the Laws of Mass Action work against it being beneficial (see Chapter 5), and if the rate of synthesis is high the cost–benefit test will be very hard to pass).

So, NP pathways that have evolved links to inducibility mechanisms will be favoured and such mechanisms will have been evolved early in evolution with little selective pressure to lose such inducibility.[17] However, from the second lesson, we learned that the next NP made by a pathway, which was formerly effective in producing an effective substance but that substance was now redundant, has a lower chance of enhancing fitness because this new substance is also chemically similar to the redundant chemical to which the target organism has evolved resistance. Furthermore, agrochemical and pharmaceutical screening programmes have proved that the value of any new chemical is unpredictable. If we apply that lesson to the evolution of NP pathways, it is reasonable to speculate that evolution will have favoured versatile inducible mechanisms, ones capable of responding to insect and pathogen attack. In effect, every new NP made as a result of mutation in an individual will be screened for its value to the producer when faced with a challenge from an insect or a fungus or any other factor linked to the inducibility system. If each induction system, initiated by insects or initiated by pathogens, was linked very specifically to particular pathways, used uniquely at one moment in evolutionary time against just a single threat, the opportunity to gain fitness by using a new NP against fungi that came from a pathway which had evolved solely to be used against insects would be very greatly reduced. Consequently, 'crosstalk' might be best seen not as an unintended, undesirable consequence of evolution but an inevitable consequence of the fact that any new NP being made has a very low probability of possessing potent specific activity against one specific kind of organism and a slightly higher probability of possessing potent specific activity against several different kinds of organisms.

What does this chapter tell us about the way science works?

The main lesson, one that has been preached before, is that the fragmentation of a subject can break up the large picture into such small pieces that the big picture is hard to see. It is as if several groups take away a random collection of jigsaw pieces, thinking that they have a complete picture. This problem not only manifests itself in the world of research but also in education systems where modularisation and the emphasis on learning 'facts' encourages fragmentation.

9

The Evolution of Metabolism

Nothing in biology makes sense except in the light of evolution.

—*Theodosius Dobzhansky*

Summary

One person looking at a road map simply looks for the shortest route from A to B while another person wonders why city A and city B were built where they were. Most biochemists looking at a metabolic map tend towards the former type of viewer, yet to really understand metabolism as a whole one needs to look beyond the roads and ponder why the networks evolved as they did. To understand metabolic networks, one needs to understand the selective pressures that operated to shape each pathway and to link the pathways together. It is the properties of new molecules, produced as a result of the mutation of existing enzymes that are the key to understanding how selection operates to hone metabolism. Three distinct properties that any one novel molecule can bring to its producer organism can be identified. An ability to integrate into the existing basic metabolism, an ability to contribute to physicochemical needs of the cell or a biomolecular contribution. The novel chemical, and the novel enzyme and the gene that cause that substance to be produced, will only be retained in the population if the novel substance enhances (or does not significantly detract from) the property mix of the organism. The selection rules that shape metabolism differ depending on the property class to which the novel substance contributes. Consequently, the metabolic traits found in different branches of metabolism will differ significantly. However, because any novel chemical can bring to the producer a benefit that is in a different property class from the substance from which it is made, the selection pressures that operate on any one pathway, in any one organism, at any one time will not be rigidly fixed. This simple model for the evolution of metabolism offers an explanation of the duality of many NP pathways discussed in Chapter 3.

In biology, many long held categorisations were finally abandoned because they were no longer productive, meaningful or they lacked an adequate evolutionary underpinning. The model for the evolution of metabolism outlined in this chapter explains why the terms 'primary metabolism' and 'secondary metabolism' should now be consigned to history.

The legacy of the split of NP research from biochemistry—evolutionary theory was not fully exploited

It is hard to appreciate the state of biological understanding before Darwin and Wallace, independently, published their revolutionary ideas in the mid-nineteenth century. Until that time, biology had largely been about gathering data and cataloguing; studies were rarely guided by theory. What was needed was a theory, which would allow the information to be assembled into a coherent story. The principles of natural selection provided a way of explaining why related organisms might differ in important respects and why unrelated organisms might share features. Biological diversity was no longer something to be described and classified, it was something to be understood and explained in terms of the basic rule of evolution—the natural selection of variants in a population such that the fittest individuals pass on their genes to subsequent generations. However, at the time when Darwin and Wallace published their work, there was no understanding of genes or indeed very little understanding of cell biology or biochemistry. The first really productive applications of evolutionary theory were explanations of the fitness of whole organisms but even then not all biologists accepted Darwin and Wallace's view that the competitive selection that honed the fitness of individuals would result in significant shifts in the fitness of the species.[1] So in the nineteenth century, the few biologists studying cell functioning were understandably still at the data gathering stage and researchers working on NPs were working in chemistry departments (Chapter 1), consequently, they were very little influenced by the exciting new evolutionary theory causing such fierce debates among biologists.

However, during the latter half of the nineteenth century, some workers began to ponder why organisms made the chemicals that were being discovered and catalogued. Those working on animals noted that the basic processes of digestion were shared by many organisms and that all higher animals seemed to have evolved similar strategies to rid themselves of 'waste'. Plant biologists also noted a commonality in the types of physiological chemistry (what we now call biochemistry) and cell biology that they were studying, for example, the green pigment chlorophyll was nearly universal in plants which had a capacity to use sunlight to capture carbon dioxide. Not surprisingly, most of these early physiological chemists worked on what seemed like the dominant biochemical pathways common to many organisms. The great, and very influential German plant physiologist, Julius Sachs (1832–1897) pronounced that plants made two kinds of chemicals; a set of fundamental chemicals common to most plants, which were necessary to survive and reproduce, and another collection of chemicals that he called 'biproducts'. Sachs, and others, had noted that 'biproducts' seemed to vary depending on the plant species and were often characteristic of individual species or even parts of individual species (e.g., the smell of flowers).[2–4] Sach's ideas were expressed more formally in 1891 by Albrecht Kössel (subsequently awarded the Nobel Prize for other work) who gave the clear binary classification of biochemistry that would remain in use for over a century.

Primary metabolites (made by *primary metabolism*)—the basic set of chemicals needed for life and the type of metabolism that is the focus of all biochemistry textbooks and most biochemistry research.

Secondary metabolites (made by *secondary metabolism*)—the chemicals made only by some species or families, chemicals that are clearly not essential for life because most organisms do not make any one of them. These chemicals include what this book has called NPs.

What is so surprising is how readily Kössel's classification was accepted, and how influential the classification was to prove. From its inception, it was clear that Kössel's classification left many chemicals made by organisms in limbo.

Lipids

Lipids are substances that are considered to be essential for cell functioning; all cells contain lipids, hence lipids must be primary metabolites? However, the spectrum of lipids produced by various species can differ very markedly; indeed the lipid spectrum of a species can help identify that species.[5] Furthermore, the study of mutants with altered lipid compositions has shown that some individual lipids can be lost from plants without any apparent loss of short-term fitness.[2] This suggests that some individual lipids are more like NPs and can be lost without a short-term loss of fitness. But clearly, cells need lipid-rich membranes; hence, some lipids are essential for life. So, lipids as a group would have to be classified by Kössel's definition of 'primary' metabolites but some individual lipids would be classified as 'secondary' (equivalent to NPs). The way to resolve this contradiction is to abandon Kössel's classification, as will be argued in more detail later, but first another instructive example.

Carotenoids

Carotenoids are a class of yellow–orange–red chemicals widely found in organisms. Most readers will be familiar with this variation in some common fruits and vegetables—tomatoes, peppers and citrus fruits all come in yellow, orange and red forms due to their different carotenoid content.[6] Clearly, this variation suggests that the individual carotenoids giving the particular colour are not essential to the fitness of that plant organ. Further, evidence comes from examining the carotenoids found in leaves, pigments that are usually hidden from our eyes by the green leaf chlorophyll. Leaves contain many carotenoids, a few in large amounts and many more in much smaller amounts. The spectrum of carotenoids in any species will be somewhat characteristic of that species. Clearly, this carotenoid variation suggests that some individual carotenoids are not essential for survival. But many experiments have shown that the total abolition of carotenoid synthesis (by using chemicals that inhibit the enzymes making carotenoids or by making mutants which lack a functional key enzyme that make carotenoids) is lethal. So, as in the case of lipids, some individual carotenoids seem to be optional ('secondary' in Kössel's classification), but overall carotenoids are essential in green plants ('primary' in Kössel's classification). Once again, Kössel's classification would place carotenoids as a group in a different classification from some individual carotenoids.

Gibberellins

The gibberellins are a group of chemicals that play a very important role as endogenous regulators in plant. These isoprenoid/terpenoid compounds (Chapter 3) have been placed in a group because they share a common key carbon skeleton. Each member of the group differs from the others on the basis of the different individual groups that have been added to that carbon skeleton. Interest in this group of chemicals began in the 1930s, when Japanese scientists were studying a fungal disease of rice that was characterised by the infected plant growing unusually tall. It was shown that the fungus was producing a chemical that stimulated the elongation of the rice plant. In the late 1940s and early 1950s, chemists in the United Kingdom and the United States continued these studies and determined the structures of several related molecules being made by the fungus (*Gibberella fujikuroi*). To distinguish the variants being studied, as each new gibberellin was isolated, it was given a number in sequence—GA_1, GA_2, GA_3, and so forth. Studies of the biological properties of members of the family revealed that some of these molecules (GA_1, GA_3, GA_4) had a high potential to cause rice elongation while others were very much less effective (GA_2, GA_5, GA_6). It was soon reported that the highly potent gibberellins were not only capable of increasing the elongation of rice but also many other species, especially if tested on dwarf forms (for instance, a dwarf garden bean would grow like a climbing garden bean if given a small dose of GA_3). In the mid-1950s, it was unexpectedly shown that plants also made gibberellins. It was soon accepted that these chemicals were endogenous plant hormones and that they played a major role in controlling plant elongation and some other important physiological responses. The search for novel gibberellins in plants intensified and over 125 have now been found, with each plant species having its own spectrum of gibberellins. As in the case of carotenoids, each species contains a few major gibberellins and a wider spectrum of minor ones. Yet within any one species, it appears that only a minority of its gibberellins possess crucial biological activity (these are clearly 'primary' by Kössel's classification) but the plant also makes some 'inactive' gibberellins (seemingly 'secondary' by Kössel's classification). So, even in the case of a group of chemicals which includes a crucially important plant growth substance, it is clear that Kössel's system of classification is problematic.

Once a classification system has clear inconsistencies, it should be time to consider whether that classification is serving any useful purpose. Remarkably, however, the classification of naturally occurring chemicals into Sachs's or Kössel's two categories not only survived for a century but was accepted largely without comment. Indeed, the terms 'primary metabolism' and 'secondary metabolism' are still widely used as any internet search will attest. Yet to ignore the inadequacy of the classification is to ignore an opportunity to increase our understanding of metabolism. At best, Kössel's classification simply gave biochemists an excuse to ignore NPs as being of secondary importance. At worst, the misconception held back our thinking about biochemical evolution for decades. This chapter takes as its starting point the view that Albrecht Kössel's classification of chemicals is no longer a useful one. It then tries to build a simple evolutionary framework that can be used to view all biochemical processes leading to the synthesis of low molecular weight chemicals.

Genes, enzymes and enzyme products—a hierarchy of selection opportunities

In every population of every species, mutations will give rise to individuals with a changed genetic makeup. The fittest individuals in that population will pass their genes to future generations more frequently, hence those genes will increase in frequency in the population. Such a simple description of evolution is commonly found in textbooks, indeed it is taught at schools. However, this description emphasises the role of the genes but it is not the genes that possess any properties that directly enhance the fitness of the organism. Genes are keepers of information but it is what that information is about that makes the difference to the organism. The genes code for proteins; hence, the fitness of an individual is really the net fitness of the mix of proteins that have been made by the individual. The genes are a way of storing information about which mix of proteins work best and it is necessary to think more about proteins when thinking about biochemical evolution. Each protein serves one of several different roles, for example, as catalysts (enzymes), structural (cell wall proteins), storage proteins (see proteins), regulatory elements or ion channels. So when a mutation occurs in a gene, such that a new protein variant is made, the outcome will depend on the role of the protein. In most cases, the functionality of the new protein will have a direct outcome on the fitness of the cell—a modified ion channel, for example, might have a direct effect on the ability of the cell to control its ionic balance. It is the properties of the new protein that directly determines the cell fitness. However, in the case of a protein that acts as enzyme, things become a bit more complex. If the mutant enzyme simply carries out the same catalytic function on the original substrate, more or less efficiently, selection will simply optimise the outcome over evolutionary time—metabolic optimisation. A more significant change would however arise if a novel enzyme activity is a consequence of the mutation, giving the mutated individual the capacity to make a novel chemical. In this case, it is *not* the properties of an individual enzyme that directly gives a mutant new properties that could potentially enhance its fitness. It is the *properties of the novel chemical*, made by the new enzyme that has the potential to change the fitness of the mutant. In other words, the mutated gene codes for a mutated enzyme but it is the properties of the chemical made by that novel enzyme that is then the focus of selection in this case. This point is so important, I will repeat it.

A gene is only as good as the protein it codes for. In the case of a gene coding for an enzyme, a gene is only as good as the properties of the chemical made by the enzyme coded for by the gene. This train of thought will soon lead us to discuss the properties of small molecules but first a short refresher on metabolism.

What shapes the metabolic map and pathways?

Biochemists have spent over 100 years isolating and characterising the hundreds of enzymes found in cells and have constructed 'metabolic maps'. Like road maps, some major highways were soon identified. Links between the highways were then sought; the byways were usually added later. But there is no universal map because organisms have evolved their own specialised biochemistry based on the chemicals available in their environment or in their food sources. A map could be constructed for any species and it would be expected to be very similar to the map of closely related species but somewhat different to the metabolic map of unrelated species and very different from the metabolic map of a species in a very different group of organisms. What these metabolic maps have in common is that they show the way in which products can be passed from one enzyme to another. In the same way that road maps are given order by arbitrary assignments of road numbers (with arbitrary start and end points), some of the sequences of transformations in metabolic maps have been given somewhat arbitrary start and end points and termed *pathways*. The problem of deciding where a pathway starts and ends is akin to the naming of roads or rivers—it is sometimes quite hard to judge where each part of a network starts and ends. Metabolic maps, like road maps, have long linear sections, branches, circular routes with no ends or beginnings. Crucially, it is sometimes possible to use multiple ways of passing between two points. However, in the same way that road maps help navigation, but are simply a human construct, so a metabolic map should be seen as something similar, a handy aid but only meaningful to a few human investigators.

Why are pathways shaped and linked in the way they are? How has evolution formed and shaped the metabolic map? These questions, like all evolutionary questions, cannot really be answered with any certainty so readers should be warned that any answers must be speculative. The crucial events happened billions of years ago and biochemistry is largely ephemeral. There is a hope that as genetic codes of more organisms are determined it may be possible to trace lineages of many individual proteins. This could allow us to increase our understanding of the evolution of proteins, a subject that has been advanced considerably during the past two decades. However, the evolution of a protein, optimising its functioning within an existing metabolic milieu, is not providing information as to how individual metabolic steps came to be assembled or incorporated into a complete working metabolism. To start the speculative process of thinking about how metabolism evolved, it is necessary to discuss some ideas about how new enzyme activities arise.

How do new enzyme activities arise?

The consequences of a mutation to biochemical processes

Mutations cause changes in the sequence of nucleotide pairs which make up an individual's DNA. Mutations are a natural phenomenon, with each gene having roughly a 1 in 1 million chance of being changed every time it is replicated. There are several types of mutation. *Base substitution* (point mutation) is where a single nucleotide in the DNA sequence is substituted for another. If that substitution occurs in a part of the DNA coding for a protein, the RNA made from the DNA might be changed with the result that a protein may be made with one changed amino acid. The effect of changing the amino acid composition of the protein can vary:

- producing a functionally useless protein (usually a lethal mutation);
- producing a protein which is functionally unchanged (selectively neutral); and
- producing a protein with properties that enhance the producing organism's fitness (a beneficial mutation).

A *frameshift mutation* causes one or more nucleotide pairs to be added or deleted and that can cause more dramatic changes in the amino acid sequences of a protein coded for by the mutated gene, with more dramatic effects on the functioning of the protein.

Because many proteins with particular functions have evolved over billions of years, there are many more chances of a deleterious mutation occurring in the genes coding for such protein than a favourable one (there are many, many more ways of messing up a protein's function than there are of making it better). Most of the favourable mutations will have happened by chance already and optimisation will have taken place. As with any evolutionary process, that 'optimisation' is limited by evolutionary history—there are only a limited number of novel positive options remaining. When thinking about an enzyme that is embedded at a position in the metabolic map there are many constraints imposed by the fact that the enzyme has evolved to serve its role *within limits set by all the properties of the enzymes before and after it in the pathway*. An analogy would be that in most countries, the road network was 'optimised' to meet local economic, social and political needs that prevailed long ago; many major cities were located because of their proximity to sea or river transport with road links developing centuries later. Those previous decisions relating to past human history limit the opportunities available today and will continue to constrain human affairs. (The predominance of major cities built at the edge of oceans might seem unfortunate in the light of predicted sea level rises!) As in the evolution of human civilisation, so in biochemical evolution, there are many more options for the equivalent of road surfacing, road widening and traffic control measures than there are opportunities for new trunk roads. The analogy can be taken further. In the same way that suddenly closing one piece of road and building a new section leading in a different direction would cause chaos, similarly it is predictable that a mutation resulting in a replacement of one enzymic activity with another would destroy the co-ordination that is the basis of metabolic co-ordination. At least in such major diversions, biochemical evolution has one answer to the problem—gene duplication. With gene duplication, as a result of a mutation an organism gains an extra copy of a gene thus allowing one copy to subsequently mutate while the other copy continues to play its original role.

New enzyme activity—it is the enzyme product not the enzymic ability that is the initial focus for selection

The current consensus is that most new enzyme activities usually arise when a duplicated gene mutates in such a way that the protein for which it codes has a changed substrate preference. This new enzyme will carry out a similar type of chemical transformation to that which was carried out by its previous form but it will act on a different substrate and hence produce a different product. How might the fitness of a cell (hence the fitness of all the higher levels of organisation—tissue, organ or individual organism) be changed by the introduction of *one* new chemical? If the mutant produces only one new chemical, it will be the *intrinsic* properties possessed by that substance and not the properties of the mutated enzyme that will be the initial focus for selection. The new substance could

- possess properties that are new and enhance the functioning of the cell;
- possess properties that are new and adversely affect the cell;
- possess properties that are new but have no impact on the functioning of the cell except in terms of any imposed metabolic cost of production; and
- possess properties that can substitute for an existing necessary property.

If the new molecule possesses *intrinsic* properties which give a cost/benefit <1 then selection will favour the retention of individuals possessing that variant. Variants with a cost/benefit >1 will be lost from the population.

However, what if the new chemical being made by the mutant can be transformed into yet another substance by an enzyme already functioning in the cell? In other words, what might happen if a mutation of one enzyme produces a new substrate for an existing enzyme? In this case, selection can act on the *intrinsic* property of the original new substance and/or on the *intrinsic* properties of next new metabolite(s). Thus when a new substance feeds into an existing metabolic matrix, the focus of selection could be on the properties of one or more *derived* compounds (Figure 9.1). However, before considering the consequences of these two different scenarios, it is necessary to consider what we mean by *the properties* of the chemicals that are made.

The three main properties of molecules that can benefit cells

Property I. Derived properties—pathways involving the basic metabolic pathways of most cells ('Basic Integrated Metabolism')

Although many think of the great diversity of chemical structures as being the defining feature of NP metabolism, maybe one should turn that idea on its head and think of the

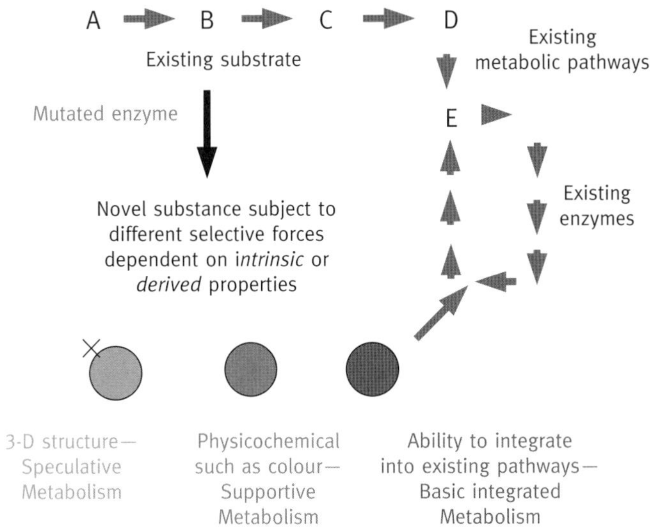

Figure 9.1. When a mutation in an individual produces a new enzymic activity, capable of producing a novel substance from an existing substrate, selection pressures will act on that individual which will be related to the cost of production of the new substance and the value of that substance to the producer. The novel substance will bring intrinsic properties to the producer, properties such as the possession of biomolecular activity or useful physicochemical properties such as colour. However, if the novel substance is converted by other existing enzymes to yet more novel chemicals, the inherent properties of those other novel substances will have been derived from the properties of the original novel substance.

lack of chemical diversity as the defining feature of the metabolism largely shared by organisms (what Kössel called 'primary metabolism')? If there is one feature that characterises such metabolism, it is the fact that most 'primary metabolites' are converted into other substances; they have to integrate into the overall metabolic network. The most important intrinsic property of any one of these substances is that it is an acceptable substrate for another enzyme; hence, the major contribution to the fitness of the producer is derived from the properties of all the other substances made from that metabolite. Two alternative ideas have been advanced to explain the evolution of 'primary metabolism'. Horowitz[6] postulated that biochemical pathways leading to the building blocks necessary for the production of structural and informational molecules (RNA, DNA) evolved 'backwards'. New enzyme variants that could introduce appropriate molecules into the evolving pathway would be highly beneficial and would be strongly selected. This would be an extreme example of selection by a *derived* trait, in that each new variant contributes to fitness by improving the efficiency of production of a substance that already possesses a useful property. An alternative model, where diverse and random biochemical transformations at some moment generated a co-ordinated function by chance,[7] is an even more extreme version of a property being *derived*—in this model, the derived property resides within the unique combination of properties of *all* the components. The important feature shared by both these models is that 'primary metabolism' would have evolved because chemical diversity was available and was being extended by chance events. Once a self-replicating, 'living' structure evolved, the main biochemical processes involved in the production of that 'living' structure would be severely constrained. A new enzyme variant arising which could produce a new molecule from a common, important precursor in a cell would be likely to impose high costs on the cell simply as a result of disrupting the flux of material from an existing important pathway. This cost would often be very high; hence, any new chemical being made would have to give very large benefits to outweigh the costs imposed by the disruption of a system that had already been improved by selection. Although gene duplication can allow the potential for extending rather than substituting chemistries, competition for substrates would have existed and there would be a high probability that such competition would be detrimental. Any new product arising as a result of mutation might also have sufficient structural similarity to an existing metabolite that might act as a substrate analogue for another enzyme, or act allosterically, both of which might have adverse effects on fitness. These types of constraints will have been very severe on all pathways through which there is a high flux and which are necessary for cell homeostasis. Because the selection pressures operating on this type of pathway are so different from those operating on pathways leading to molecules selected on the basis of their intrinsic properties, it is predictable that metabolic traits will differ from those found in pathways leading to chemicals selected for their intrinsic properties. For example, the high substrate specificity is predictable in enzymes producing chemicals where only derived properties are the focus of selection. One can only gain effectively from derived properties if the path leading to those derived benefits is efficient.

Property II. Specific physicochemical properties—pathways leading to chemicals with a beneficial physicochemical property ('Supportive Metabolism')

When chemists began to isolate and characterise the chemicals found in organisms, they often grouped chemicals sharing similar physicochemical[8] properties into broad groups—lipids, carotenoids, flavonoids, pectins, hemicelluloses, polysaccharides and phenols. It was the shared physicochemical properties of a group of chemicals that enabled them to be extracted or quantified as a broad group. For example, lipids, which are usually highly non-polar, are easily extracted from cells by non-polar solvents such as chloroform but are not extracted by polar solvents such as water. Like dissolves like. The selective extraction of one whole group of chemicals with a common physicochemical property is still commonly used as the first stage of isolating a particular naturally occurring chemical from living tissues.

Such extractions of broad categories of chemicals are selective between major groups but unselective in terms of individual molecules within the group. Consequently, once a group of chemicals has been selectively extracted on the basis of the shared physicochemical properties, techniques of analysis that depend on specific, or more refined, physicochemical properties (such as chromatography) are used to separate the chemicals within the group and the diversity of chemicals within the class is revealed.

Why does one organism make such a diversity of lipids (or carotenoids or polysaccharides, etc.)? Surely if many lipids possess similar properties, evolution would, on cost saving grounds, favour organisms that made the minimum number of types necessary? One might have expected that, at a very early stage in the evolution of microbial life, competition had selected an optimum mix of lipids for making a membrane with a honing to perfection rather than further diversification? The answer to these questions is that it is a physicochemical property that is being selected for by evolution and because that property is only loosely linked to the detailed fine structure of the molecule a wide tolerance for variants will exist. Consider the diversity of coloured pigments in plants and some microbes.[9] If a carotenoid pigment molecule is made by a plant, variants of that molecule that differ in parts of the structure that do not very greatly influence the colour of the molecule, but which share with the original pigment the part of the structure that gives the molecule its colour properties, will be just as good at producing the colour that benefits the plant (Figure 9.2). Consequently, when a chemical variant arises within one of these broad classes, there is a reasonable probability that the variant will possess similar physicochemical properties to that of the product which had been made by the original enzyme (this is in stark contrast to the very low probability of the chemicals possessing similar biomolecular properties). When the new pigment has been produced by the mutation of the duplicated gene, it is possible that the new and the old products will share similar physicochemical properties and there is a reasonable probability that the mutation will be selectively neutral. It is therefore predictable that if certain types of physicochemical property are useful to cells, but the property is not highly specific to a particular structure, a diversity of chemical types will be found within a single organism and different organisms will tend to possess a different mix of appropriate chemicals.

This line of reasoning can be taken a little further because it provides yet another example of why, once a new biochemical pathway exists in a population, evolution will not automatically hone the enzymes to be increasingly substrate specific. Suppose one has two individuals in a population. One individual has three enzymes, each substrate specific hence making three coloured products. It is competing with another individual which makes three equivalent enzymes which possess broader substrate tolerances hence this individual makes more than three coloured products. It is not clear why the individual with narrower substrate specificity would be fitter; hence, there might be little selection pressure under such circumstances to drive the evolution of the enzymes to become more substrate specific. The fact that enzymes with broad substrate tolerance, working in matrices, have been found[10] (Figure 9.3) to produce a mix of carotenoids would indeed suggest that evolution does not always result in increased substrate specificity of an enzyme. There may also be an advantage to an individual organism in

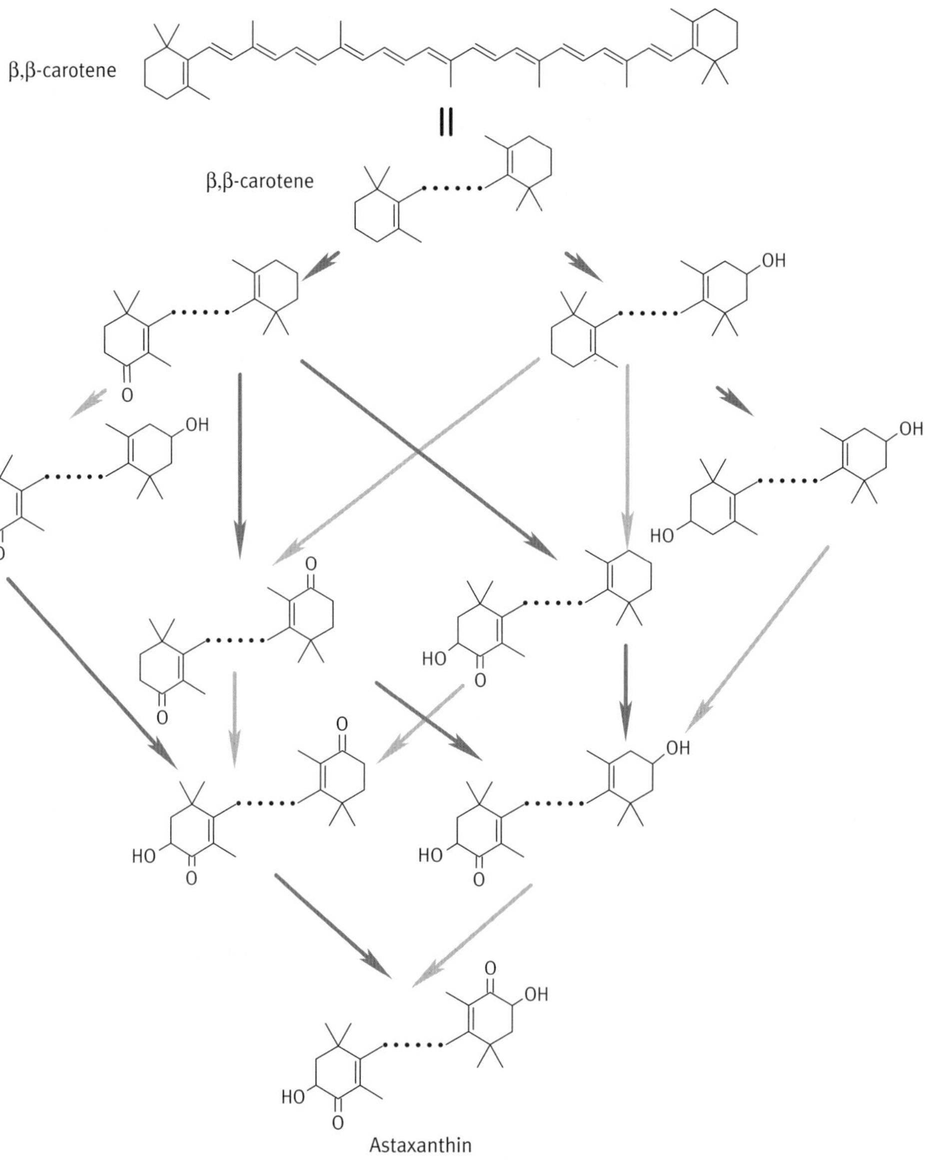

Figure 9.2. The inherent metabolic flexibility of the isoprenoid pathway leading to the synthesis of some carotenoid pigments. Genes coding for two enzymes capable of acting on carotenoid structures were introduced into *Escherichia coli* which had already been transformed to give it the capacity to make β,β-carotene. Both of the two introduced new enzymes (one shown with red arrows and the other with blue arrows) acted on multiple substrates because of their lack of specificity. The resulting matrix of transformations means that nine different products can be made by just two 'tailoring' enzymes. (Adapted from Umeno et al.[10] who used data from Misawa et al.[10])

producing such chemical diversity if the chemicals made play a role in excluding other organisms from the cell or organism (e.g., the cell wall, the cuticle). For example, microorganisms seeking to invade plant cells have to degrade the cell wall. Consequently, a chemically diverse wall would be expected to be less susceptible to degradation than a chemically homogeneous one.

One can use the same logic which explains carotenoid diversity to explain lipid diversity. Lipids that are found in membranes contribute their individual properties to the overall properties of the membrane. It is the collective net properties of the membrane that will be the focus for selection and the value of any one new lipid structure will depend to a degree on how complementary it is to other lipids being made in that membrane. Lipids are a particularly interesting case because in organisms that do not regulate their temperature, the functioning of the membrane will be influenced by the fluctuating temperature; hence, there will be an optimisation over long and short time scales. This would suggest that for many organisms there will be no single lipid composition that will be clearly optimal and different mixes might be evolutionarily indistinguishable. This logic would predict a tolerance of individual lipid composition and little selection pressure to drive enzymes making all lipids to be highly substrate specific.

There might also be another reason why chemical diversity is retained among groups of chemicals made to enhance the physicochemical properties of a cell. The chemical diversity retained in the groups of chemicals retained for their physicochemical properties would be a valuable resource as a pool of chemical diversity to be drawn upon for the generation of new compounds potentially possessing potent biomolecular activity. There is evidence that is consistent with this concept.[1] The internal cell regulators IP_3 and diacylglycerol are derived from a lipid as are prostaglandins and the jasmonates. The carotenoid pathway (isoprenoid/terpenoid) serves to provide the precursor of the plant 'hormone' abscisic acid and the fungal mating substance trisporic acid. There are numerous examples of small molecules derived from cell walls possessing biological activity which may be important in plant–microbe interactions.

Property III. Biomolecular activity—pathways leading to physiologically active compounds ('Speculative Metabolism')

As discussed in Chapter 5, because of the nature of protein–ligand interactions, it is necessary for a chemical to have a structure that precisely fits the binding site on the protein with which it interacts. The fit must be precise enough to give a binding affinity for the protein–chemical interaction such that a significant occupation of the binding site occurs even when the chemical is present at very low concentrations. These are very strict constraints consequently very few chemicals will possess the appropriate structure to bind to a protein when both the chemical and the protein are present at low concentrations. This low probability of any chemical possessing potent biomolecular activity must have been a severe evolutionary constraint on the ability of an organism to gain fitness by producing chemicals with potent biomolecular activity. Consequently, the ability to generate new chemical diversity is a trait that would have been selected for in organisms making such chemicals. The metabolic traits that encouraged the retention of existing chemical diversity, even in the absence of a current role for some products, would have also been selected. The predicted metabolic traits, predicted by the Screening Hypothesis, were discussed in Chapter 5. It might be timely now to look again at Figure 3.2 where some of the general lessons of NP biosynthesis were summarised. The concepts shown in that figure were not clearly known when the Screening Hypothesis was proposed, yet it is striking that the general rule for NP biosynthesis is that the early stage (Phase 1) of each of the major pathways incorporate stages of generating many alternative carbon skeletons and the later phase (Phase 2) provides a means of modifying the many different skeletons in similar ways to produce a myriad of different chemicals.

Why are there so few major NP pathways?

In Chapter 3, it was noted that remarkably few biochemical pathways lead from Basic Integrated Metabolism into the pathways that serve to produce substances with Type II and Type III properties. Using the concepts discussed, it is possible to offer one evolutionary scenario to explain why a few, rather than many pathways flow from Basic Integrated Metabolism. As in the case of most evolutionary arguments, it is not difficult to produce a credible scenario but hard to find supporting evidence simply because the postulated events would have occurred early in the evolution of life and there are no known 'biochemical fossils'. However, some computer modelling has at least provided some support for the approach adopted.

Consider two extremely different outcomes of generating chemical diversity (Figure 9.3). Let us suppose that billions of years ago, as life emerged, there was a rapid canalisation[11] of Basic Integrated Metabolism and that a few pathways led to substances with necessary Type II properties began to form (lipids for membrane and later pigments to provide ultra violet (UV) screening and protection from photo-oxidation). At this early stage of the evolution of cells, the different selection pressures operating on the Type I and Type II pathways would begin to shape metabolism. In Type I metabolism there would be a tendency to reduce chemical diversity to a minimum, while in Type II metabolism there would be less selection to reduce chemical diversity. However, once the basic physicochemical needs of the cell were met, there would be little advantage for a cell that mutated to start yet another branch from Basic Integrated Metabolism. Any new branch from Basic Integrated Metabolism would inevitably disrupt carbon flow in those fundamental pathways, and an organism that had already evolved to optimise Basic Integrated Metabolism in response to the existing carbon flow into Type II chemicals would have a high chance of being less fit with yet another branch from Basic Integrated Metabolism. Now let us add to this scenario the possibility that a mutation to a Type II pathway produces a chemical

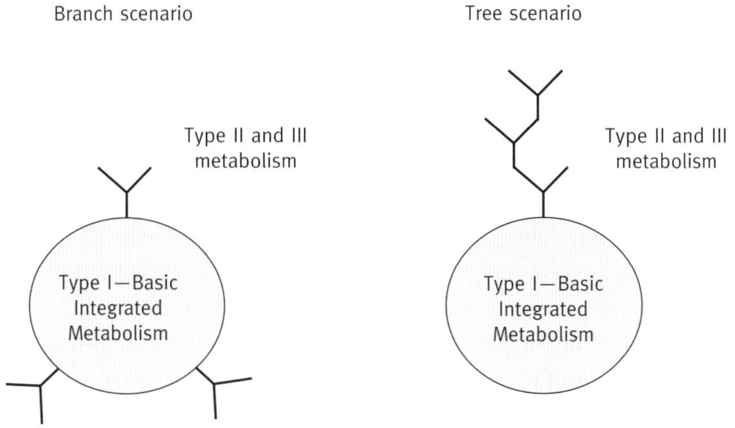

Figure 9.3. Two extreme scenarios for the evolution of Type II (Supportive Metabolism) and Type III (Speculative Metabolism) chemicals from the Basic Integrated Metabolism (Type I). On the left, mutations lead to branches from Basic Integrated Metabolism, branches that lead to the production of new substances but each new pathway is subsequently lost when the substances they produce become redundant. On the right, mutations lead to branches from Basic Integrated Metabolism, branches that are extended to produce new substances which are increasingly unlike Basic Integrated Metabolites. The scenario on the right is closest to reality, possibly for reasons discussed in the text.

which just happens to possess beneficial Type III properties—beneficial biomolecular properties. Once again, it seems likely, for the reasons already given, that Type III chemicals would more likely to evolve from a Type II pathway than a Type I one. Indeed because an organism making a chemical closely related to those substances being used in Basic Integrated Metabolism is more likely to suffer a loss of fitness due to some interference with Basic Integrated Metabolism (because new chemicals related to common metabolites are more likely to interfere with Basic Integrated Metabolism than chemicals that are less closely related structurally), it is predictable that chemical diversification near the ends of the Type II pathways would be favoured compared to chemical diversification at steps nearest the Basic Integrated Metabolism pathways. This line of argument predicts just the kind of shaping of metabolism leading to Type II and Type III substances that has been described in Chapter 3. Indeed when a computer simulation was built,[12] assigning probabilities of any new substance generated by mutation at any position in an emerging metabolic pathway having certain beneficial or detrimental properties and accounting for the carbon cost of production, the simulation revealed that one could account for the observed shape of metabolism. The model with a few pathways leading from Basic Integrated Metabolism, but with chemical diversity branching out like branches on a tree from those trunks, was favoured.

Primary and secondary metabolism—outmoded terms?

Given that three different properties (and there could be more) have been identified that govern the selection process that shape metabolic pathways, it is inevitable that Sach's and Kössel's binary classification would be problematic. The key idea introduced in this chapter is that more attention needs to be given to the role of individual molecules in cells because it is the properties of each and every chemical, not the properties of the enzyme(s) making them, that are the initial focus of selection leading to changes in metabolic capacity. Three very different roles for endogenous chemicals in cells have been introduced in this chapter and it has been argued that the very different properties of the chemicals which fall into each category have led to somewhat different evolutionary selection pressures operating in each type of metabolism. With these ideas about the importance of recognising that new enzymes are only retained in a population if the chemicals those enzymes make contribute one or more of the three classes of property that enhance the fitness of the producer, one can begin to appreciate why Kössel's categorisation was inadequate. Thinking in terms of the properties of molecules, rather than chemical structures, allows one to predict that every organism would be expected to possess a collection of molecules with the appropriate properties but it is the properties that are needed, not specific molecules. Consequently, one expects that evolution might have caused a radiation within a category, such that different molecules might play similar roles in different organisms because they share similar properties (chitin vs. cellulose; starch vs. inulin; etc.). Likewise, similar chemicals may play different roles in organisms because each can possess more than one property (flavonoids acting as UV screens and as signalling molecules). The recognition that it is the properties of individual molecules that determine the value of a pathway, and that a pathway can contribute to different property classes, reveals the inadequacy of the old classification of primary and secondary metabolism. The old classification, which should now be seen as outdated operational convenience rather than being based on any principle, was especially poor at explaining why many important chemical groups (lipids, carotenoids, polysaccharides, cuticular waxes, etc.) were so variable between organisms and why chemical diversity existed so widely within a group. For instance, plants must be able to make lipids to survive yet it is clear that not every lipid is essential. This contradiction can now be seen to be due to the fact that most lipids are selected because of their physicochemical properties. The pathways leading to lipid synthesis are essential, but there is only a limited selection to constrain lipid structure because the properties selected for are possessed by a wide range of structures. The recognition that the properties of molecules are the main focus of selection during biochemical evolution and that similar chemical structures may be of benefit to an organism because they possess different beneficial properties allows one to appreciate that any simple broad classification will fail. The pathways of 'primary metabolism' are simply pathways where there has been a very strong and ancient 'canalisation',[11] and selection has operated on the individual enzymes to ensure optimal functioning with respect to the overall efficiency of the pathway and the co-ordination of that pathway with other essential pathways. The pathways leading to NPs are pathways (Type III—'Speculative Metabolism') where a rare, sometimes ephemeral, property ('biological activity') is being selected for and the selection pressures will be different, or applied to a different degree, from those operating on the essential metabolism (Type I—Basic Integrated Metabolism). Furthermore, metabolism will not be static in evolutionary terms and selection pressures will change as metabolism matures. A pathway evolved initially on the basis of generating compounds with intrinsic properties might gradually become one which has some properties of a pathway that is selected for on the basis of derived properties. The evolution of a molecule with biomolecular activity, which acts on the producing organism, generates a quite different selection pressure compared to the production of a chemical that acts on another organism. These complexities suggest to us that it is time to start discussing metabolism as a single subject which encompasses the biosynthesis of all chemical structures. The lack of a theoretical basis for the splitting of metabolism into 'primary' and 'secondary' should be confronted and a more robust evolutionary framework developed. The differences in various biosynthetic pathways are not that they follow different rules, rather they apply the same rules to a different extent because they operate with different evolutionary constraints. These rules and constraints must be understood if a full understanding of metabolism is to be achieved and if attempts to control or change metabolism in organisms are to be successful.

What does this chapter tell us about how science works?

> *A long habit of not thinking a thing wrong gives it a superficial appearance of being right.*
>
> —*Thomas Paine.*

Paine, a great radical thinker, observed that most humans are basically conservative. This conservatism can be seen in science. It is a remarkable fact that the fundamental binary categorisation of naturally made chemicals, made 100–150 years ago, is still widely accepted despite the deficiencies discussed. The lack of concern about these deficiencies might have been excusable 100–50 years ago because of the schism of the subject that is now called biochemistry (see Chapter 1). However, it is no longer excusable and the complacency is remarkable. The subject of genomics, so solidly built on an evolutionary perspective, supports the subject of proteomics, another research area with a strong evolutionary base. But the complementary subject of metabolomics clearly has a very inadequate evolutionary basis. While the evolutionary model of metabolism[1] discussed in this chapter is new, and consequently might not have a long-term value, surely all biochemists should at least acknowledge that it is time to abandon the use and teaching of Kössel's classification and his terminology. To continue to teach something that is inconsistent with evidence would be nothing new; but it would still be inexcusable.

10

The Genetic Modification of NP Pathways—Possible Opportunities and Possible Pitfalls

If no one ever took risks, Michaelangelo would have painted the Sistine floor.

—*Neil Simon*

Summary

NPs are such an important part of the world's economy that it was inevitable that academic and industrial scientists would cast their eyes over the organisms making NPs and consider how they might usefully, profitably or interestingly modify those organisms by changing their genetic composition. However, because of the metabolic traits of pathways leading to NPs (see Chapters 5 and 9), it is predictable that the manipulation of these pathways will sometimes give unpredictable outcomes. Current methods of evaluating the safety of *genetically modified* (GM) crops are not well suited for judging the risks of intentionally or unintentionally manipulating NPs. Fortunately, because there is a low probability that any new substances being made by a manipulated organism will possess potent biomolecular activity, the risks will usually, but not automatically, be small.

What is genetic manipulation?

This is not an appropriate place to discuss exactly what the genetic manipulation of an organism is or how it is achieved. There are many books and websites that will explain the process more elegantly and authoritatively. Suffice to say that scientists now have the ability to change the genetic code of an organism, changing the sequence of nucleotide bases so that a new genetic variant is created. The extent of the manipulation varies very considerably, depending on the aim of the exercise and how well it has been carried out. The changes range from minor to major and could be one or more of the following:

- The organism can be genetically modified to make a minor variant of a protein that it normally makes.

- The organism can be genetically modified to make a protein it normally makes in larger or smaller quantities, or in different cells, tissues or organs or at different times or in response to different stimuli.
- The organism can be genetically modified so that it now makes one or more exotic proteins, those proteins were made previously only by other species.

The commonest manipulation is the insertion of one or more new, exotic genes into the DNA of an organism. Although the term genetic engineering is sometimes used to describe the process, the current state of the art is more akin to engineering as practied by a nineteenth-century blacksmith than a twentieth-century aerospace engineer. The location of the inserted gene is usually random and often unknown. Genes that get inserted into sequences that are essential for short-term survivorship of the recipient kill the organisms, so the few organisms that survive the insertion technique must incorporate the new sequences into a less vital part of the genome. Most techniques used to do the insertion are very inefficient, hence it is common practice to insert at least two genes: one gene coding for the ability of the recipient to resist a toxic substance and the other gene coding for the important gene one wants to insert. If the manipulated organism survives when subsequently exposed to the toxin, it must have incorporated the gene coding for resistance to the toxin into its genome and hopefully the other important gene one seeks to insert will also be incorporated.

Why might one want to genetically modify an organism to change its NP composition?

There are two reasons why one might want to genetically modify the NP composition of a plant or a microbe. The first reason is an academic one. One might want to change the NP composition in order to judge the consequences of carrying out that manipulation when seeking to understand the synthesis or role of a particular NP or group of NPs. The second reason is a commercial one, driven by a desire to change the NP composition of a plant simply to gain increased value. It is this second type of work that is the focus of this chapter.

How might one increase the value of a plant by changing its NP composition?

There are so many different NPs, made by so many different organisms and used in so many ways that it is inevitable that many different goals will be identified but these will usually be based on some generic approaches such as

- enhancing the amount of an existing NP in a species that already make that product—for example, increasing the accumulation of a taxol precursor (see Chapter 7);
- changing where in an organism that an NP is made in order to reduce the cost of recovery;

- changing the relative amount of some NPs so that some increase and some decrease—diverting the flow of carbon between shared pathways to increase desirable NPs and reduce the synthesis of less useful NPs;
- changing the type of NP being made;
- enhancing or changing the flavour, for example, increasing the chemicals that give an apple a Cox's Orange Pippin flavour so that a poorly flavoured apple variety with a high yield becomes more valuable;
- enhancing or changing the odour, for example, giving a pretty but odourless rose the wonderful rich rose scent of the variety Fragrant Cloud;
- enhancing NP-linked disease resistance, for example, making all gooseberry varieties as resistant to mildew attack as the variety Careless;
- enhancing the NP-linked pest resistance, for example, making cotton plants resistant to cotton boll worms;
- improving the nutritional quality, as exampled by the attempts to make rice produce more carotenoids to enhance vitamin A in the diet of poor consumers in the Far East;
- enhancing or changing the colour of an organ, for example, producing carrots with shades of yellow to intense red, by changing their carotenoid composition;
- enhancing the NPs used by parasites to locate their prey, for example, enhancing the production of volatile chemicals that parasitic insects of common plant pests use to home in on their targets, hence promoting a more effective biological control of the insect pest.

How might an understanding of the Screening Hypothesis inform attempts to manipulate NP composition?

Predictably unpredictable

If an enzyme involved in NP synthesis is introduced into another organism, an organism with its own NP profile, there is a reasonably high probability that the introduced enzyme will act on more than one substrate to give more than one product (see Chapters 5 and 9). Consequently, there is an inherent unpredictability of the outcome of genetically modifying pathways that contribute to NP diversity. The organism making NPs has evolved to give uncertain outcomes—the unpredictability is built-in and nothing to do with the unpredictability of the actual genetic manipulation process. So it must be recognised that the experience gained by studying any one example of the genetic manipulation of a plant is inevitably specific to that specific example.[1]

Adding a gene to supplement NP synthesis

If one combines the classical 'One Gene—One Enzyme' hypothesis, which won Beadle and Tatum the Nobel Prize in 1958, with the generally accepted view of most biochemists that every enzyme has evolved to convert one substrate to one product, it seems logical to conclude that the addition of one gene will add one new product to a cell. However, as discussed in Chapters 5 and 9, this simple view of biochemical engineering, a view that prevailed at the time that genetic manipulation of organisms was being first attempted, is too simplistic.

As explained in Chapter 9, there will be pathways where evolution will be favouring the reduction of uncertainty and pathways where flexibility and uncertainty might be selected for, or certainly not selected against. Consequently, the addition of a gene coding for an exotic enzyme into an organism must inevitably carry with it a probability of an uncertain outcome.[2] A detailed knowledge of the properties of the enzyme in its native organism is only partly useful because it is the properties of the enzyme in its new biochemical environment that will determine which chemicals it transforms and at what rate. This problem is most acute when manipulating pathways involved in NP synthesis because it is already known that single gene mutations in enzymes involved in such pathways can result in multiple, sometimes unexpected, changes in chemical composition. Remember the example of the spearmint mutant that became similar to peppermint that was discussed in Chapter 5? This was an example of how a *natural* mutation of a single gene gave rise to a very dramatic and significant change in the NP composition of a plant. The gene coding for the one enzyme that switched the type of monoterpenes from spearmint type to peppermint type could be isolated. That gene could be added to another type of spearmint plant to give a plant that in theory would now make both spearmint oils and peppermint oils. But we know that the new gene added to the spearmint, the gene that codes for one enzyme, will have caused several new products to be made. It is also possible that by making a plant which expresses both these genes, some more new quite unexpected products will appear because there will be more substrates than ever in the new plant. Furthermore, the carbon flow into the new and existing pathways will be unpredictable so the relative composition of the NPs that will be found will be unknown. So tinkering with NP pathways is inevitably going to be unpredictable—the types and the quantities of the NPs may or may not change after a new NP enzyme coding gene is introduced and expressed. This could be an iterative process. One gene added to a plant or microbe with a rich NP profile could in theory produce many new chemicals, each with unknown properties. This prediction has already been experimentally verified. A gene coding for (*S*)-linalool synthase, taken from *Clarkia breweri*, was expressed in three different plant species, tomato, petunia and carnation.[3] Each of these different species made the expected *S*-linalool from their own endogenous geranyl diphosphate but tomato also made 8-hydroxylinalool, petunia also made linalool glycoside and carnation made two linalool oxides (Figure 10.1). In other words, the existing NP metabolic flexibility in these three species further elaborated the expected novel substances. So unlike the uncertainty associated with Bt gene insertion, where the uncertainty in outcome lay at the ecological level, when one alters the NP composition of an organism you have uncertainty at the biochemical level and even greater uncertainty at the ecological level because NPs are so important in determining the interactions between organisms.

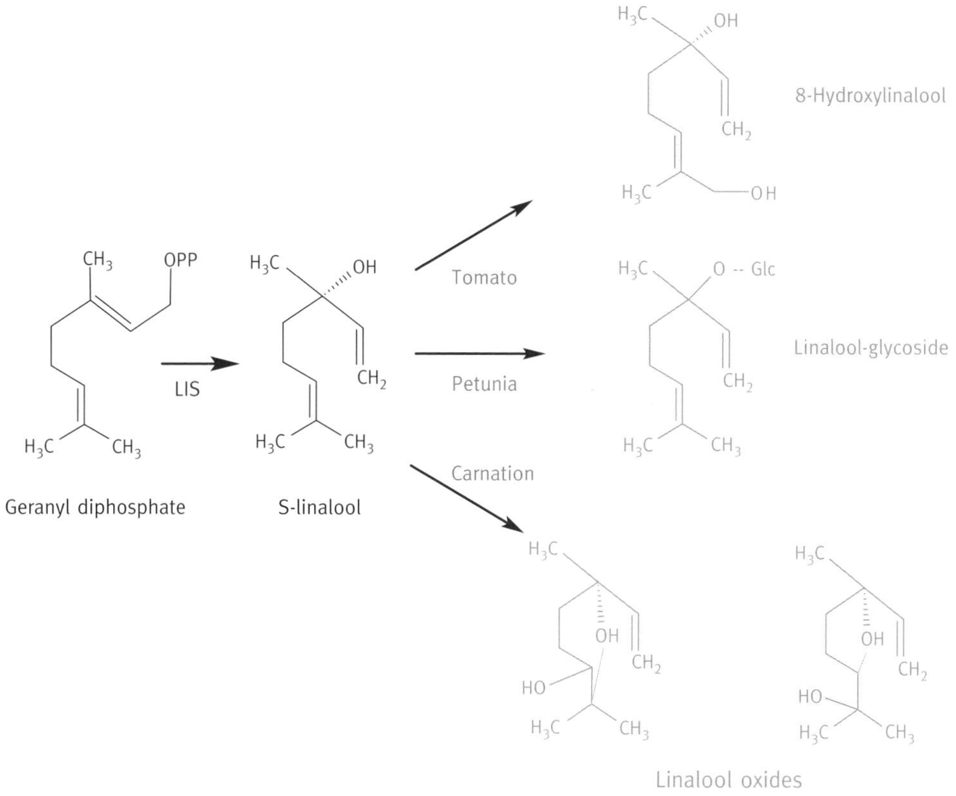

Figure 10.1. The gene from *Clarkia breweri* coding for (*S*)-linalool synthase (LIS) was added to three different plant species (tomato, petunia and carnation) and each species produced the expected product, *S*-linalool. However, the existing NP metabolic flexibility in each species allowed the novel substance, *S*-linalool, to be converted to other substances, those substances being different in each species due to the differences in NP metabolism in each plant.[3]

Evidence for certainty

Not everyone accepts that the genetic manipulation of plants is unpredictable. Kutchan[4] concluded that plants can be tailored in a rational manner with marginal effects and hailed the work of Kristensen et al.,[5] as being a milestone in the public acceptance of genetically modified plants. The elegant studies of Kristensen et al. showed that it was possible to add genes coding for enzymes responsible for the synthesis of an exotic NP (dhurrin) to a plant (*Arabidopsis thaliana*) with no evident developmental or morphological consequences and only very minor changes in the chemical composition. This finding would seem to counter the argument advanced some years ago[6] and summarised above. However, Kristensen et al. added a new functional metabolon (a group of enzymes spatially oriented in respect to each other) and this inevitably reduced the opportunity for inherently promiscuous enzymes to act on the exotic new intermediates. Such metabolic channelling of some stages in secondary product metabolism may well be the result of evolutionary selection tempering the inherent capacity of secondary metabolism to generate chemical diversity. However, there is evidence that such channelling is not universal.[7] It is possible to speculate that the advantage of evolutionary selection favouring the metabolom strategy to reduce the impact of enzyme promiscuity, rather than the alternative strategy of tightening the substrate specificity of the individual enzymes, is that a greater capacity for promiscuity can be retained and released by subsequent mutations. Indeed, such 'hidden pathways' were predicted as part of the Screening Hypothesis (see Chapter 5). Consequently, the fact that one part of an exotic pathway can be inserted into a plant with predictable results by no means provides a universal lesson.

In summary, both experimental evidence and the evolutionary model suggest that the manipulation of NP pathways will often produce unexpected changes in NP composition. Such manipulation will be predictably unpredictable. But can this unpredictability be compensated for by a more thorough study of the new NP composition?

Metabolomics—what it can and cannot tell us

The term metabolomics is a recent one, a term introduced after the terms *genomics* and *proteomics* became fashionable. *Genomics* was the generalised term used to encompass the knowledge that comes from identifying the genes that occur in an organism. Given that genes codes for proteins, the term *proteomics* was introduced to cover the methods, of identifying and quantifying the proteins that are made in an organism. While *genomics* were something quite new, *proteomics* was really a rebranding of a much older and well-established subject—the study of proteins and their contribution to cell functioning. However, by using the fashionable suffix 'omics', the subject could claim to be a part of the 'new biology' that attracted so much funding at the end of the twentieth century. However, the contribution of one class of proteins, the enzymes, to the current status of the cell, was not easy to judge simply by their presence or absence. It was known that the presence of an enzyme protein in a cell did not reliably predict whether it was currently active. A number of ways were known of regulating the activity of an enzyme, many of which were highly dynamic (e.g., feedback inhibition). Thus although *proteomics* could address some of the unknowns that *genomics* could not, uncertainties remained when judging the actual metabolic functioning of a cell. Recognising that the contribution of enzymes to the current status of a cell could possibly be best judged by measuring the products enzymes make, some analytical chemists and biochemists rebranded their subject and *metabolomics* was born. *Metabolomics* is the study of the *metabolome*; the metabolome is the complement of all the small molecules in an organism. While the term metabolome is fashionably recent, it is misleading to claim that that metabolomics is a new discipline. The concept of analysing the chemical composition of organisms stretches back at least

200 years, as summarised in Chapter 1. Clearly, the renewed interest in the chemical composition of plants and microbes is to be welcomed but there needs to be a caution as to exactly what such an approach can deliver.

There are some questions that need to be answered concerning the ability to fully describe the metabolome of any organism:

- How easy will it be to complete a full analysis of the metabolome of any organism, let alone a genetically manipulated variant?
- How can that information inform us about the risk that the organism presents to organisms that interact with it (humans and other organisms if a plant or a microbe is grown in an open system)?

The challenge of conducting a complete chemical analysis

It is a remarkable fact that no complete chemical analysis of any plant has been published. The full genome analyses of several species of plant are available in publicly accessible databases and the protein composition of these plants can be partly predicted from these data. However, the chemical composition of even important crop plants has not been fully explored. Why?

The general public, and even many undergraduates studying science, often underestimate the difficulty of carrying out an analysis of chemicals in a sample. Given a sample to analyse for a 'poison', the chemist will normally ask several questions:

- Which chemical(s) do you want to measure? Every chemical needs its own method of analysis—it is the unique properties of an individual chemical which allows the chemist to find it among the thousands of chemicals also likely to be present in the sample. If the chemical being sought is a synthetic one, which of the 80,000 chemicals made by humans is to be sought?
- At what level of sensitivity do you want the analysis to be conducted? The difficulty, hence the cost, of carrying out an analysis rises as the sensitivity needed increases.

Thus the chemist given a sample to analyse, and with only a limited budget, will either have to analyse a few chemicals very sensitively or a wider range of chemicals with less sensitivity. The effort required to conduct a thorough analysis will depend to a large extent on the information already existing about the chemical composition of the sample.

The simplest case to consider would be one where a plant had been genetically modified to make a new protein, a protein without enzyme activity, such as the introduction of the Bt toxin.[1] In this case, the concept of *equivalence*[8] is typically applied. The case is made that the chemistry of the unmodified plant and modified plant are likely to be so similar that they can be considered to be equivalent.

Clearly the '*primary metabolites*' (what were called *basic integrated metabolites* in Chapter 9) are the easiest types of chemicals to be analysed because they occur in highest concentration.[9] Because so much of primary metabolism is shared by plants, it is not unrealistic to expect that methods will become available that can routinely, and largely automatically, report the concentration of the several hundreds of primary metabolites in a sample. In a genetically modified organism with an altered basic metabolism, it is also to be expected that many significant changes will have already revealed themselves by changes to the development, morphology or growth rate of the organism. However, even such dramatic changes might be hard to interpret in plants because the plasticity of plant development will enable a small localised change to be propagated into larger ones as a result of alternative developmental pathways opening up for the whole plant. Fortunately, there is a considerable body of knowledge available to judge the effect of changing the concentration of certain key metabolites on the well being of an organism. However, because different organs, or indeed different cells, at different times, under different conditions will have very different metabolite concentrations, there can be no universal 'metabolomic analysis' for even a single organism. Thus, the tools that facilitate the analysis will need sensible and considered use with the limitations and uncertainty of the analysis given some prominence.

The difficulties in providing a full and understandable analysis of primary metabolites are small compared to the problems that face those seeking to show that the genetic manipulation of NPs may or may not be of consequence. NP chemicals will inevitably be much harder to analyse because every plant and microbial species will possess a unique spectrum of chemicals. Hence, unlike the methodologies being developed for the analysis of 'primary metabolites', the specific methodologies needed for a thorough analysis of the NP composition of one plant species might be only useful for that species and its close relatives. Furthermore, because NP metabolism is predictably unpredictable, an organism expressing an exotic gene coding for an enzyme involved in an NP biosynthetic pathway might be producing several unknown new structures. Looking for known chemicals, for which a methodology has been painstakingly developed, is hard but seeking unknowns is a much bigger challenge. Determining the structures of these new, possibly rare chemicals might require larger quantities to be extracted and ultimately confirmation of the chemical structure might require chemical synthesis—a huge undertaking for many NPs. Thus whilst the metabolomic analysis of primary metabolites might be built on a database with 1000 known primary metabolites, a metabolomic analysis of NPs might need a database 100-fold to 1000-fold larger—with the majority of that data currently unavailable. Furthermore, at what level of sensitivity should the analysis be conducted? The common experience of those analysing NPs is that if one increases the sensitivity of the analysis, the number of compounds which reveal themselves increases significantly. Given that most NPs found in an organism would be expected to play no significant role in increasing fitness of the organisms making them (see Chapter 5), it is tempting to suppose that the chemicals made in the largest amounts must be most important to the maker (for good cost–benefit reasons) but that can only be an assumption (see also Figure 4.5).[10] To further complicate the picture, it is known that the NP composition of an organism is very greatly influenced by the conditions under which the organism is grown.[11]

What can we deduce from the analytical data?

As outlined above, it is predictable that any significant change to 'primary metabolism' will very often result in a deleterious effect on the plant or microbe, hence are unlikely to be of commercial value. Furthermore, the majority of 'primary metabolites' are unlikely to pose a threat to those who consume them. Most generalist organisms that consume plants or microbes have evolved the capacity to metabolise these chemicals, indeed their survival depends on the ingestion of these chemicals, and it is normal for such organisms to vary the mix of these primary metabolites on an hourly, daily or seasonal basis. These consumers are likely to have evolved methods to tolerate large changes in the concentration of primary metabolites in their diet. Hence, a metabolomic analysis of primary metabolites is not easy to justify on the grounds of human food safety but it could be more important in terms of judging any undesired effects on other consumers of the genetically manipulated product. For example, many insects are highly specialised herbivores and will have evolved with a very consistent diet and hence may not have a capacity to tolerate changes in the primary metabolite composition of their diet without a loss of fitness.

What might a metabolomic analysis of NPs of a genetically manipulated plant tell us about the wisdom of adopting the widespread cultivation of such a crop? This question cannot be answered in general terms because there will be so many unknowns and/or assumptions involved in producing an answer. In contrast to the case of plants with changed 'primary metabolite' composition, where there are theoretical reasons to accept that the majority of consumers of the products will be preadapted to tolerate all but very large changes in 'primary metabolite' composition, in the case of changes in the composition of NP one cannot make any assumptions that the consumers will be preadapted.

Lets us consider, as an example, a genetically manipulated plant that has been found by metabolomic analysis to produce three novel NPs in small amounts—say 5% of the mass of the major NP normally found in that species. What understanding does this new piece of information give us in respect of the safety of this crop for humans or for other members of the natural world? There is a very high probability that these novel chemicals will have completely unknown properties; consequently, it will be impossible to say whether these chemicals pose a risk to any organism that comes into contact with the plant. The Screening Hypothesis predicts that the probability of any one of these chemicals possessing potent, specific biological activity (or more accurately biomolecular activity) is very low. In other words, at this stage of the analysis, the actual identification of the new chemicals offers little more reassurance that the theoretical underpinning of the subject overall. For the evidence to surpass the theoretical logic, precise toxicological studies of the new chemicals would be needed. To undertake such studies would require larger quantities of the new chemicals to be made or extracted. This in itself would be a considerable task if these chemicals occur at low concentrations or if these chemicals are very difficult to make in the laboratory (which many NPs are). Even if such studies were undertaken, given that similar toxicological data will be unavailable for the great majority of NPs that occur in the same plant, there would be no appropriate reference point to use to judge whether the risks to consumers (human or otherwise) of the genetically manipulated plant would be greater, or less, than the original plant.

A further problem presents itself in that the NP composition of a plant varies significantly depending on the challenges that the plant has experienced or is experiencing. Temperature, water, insect infestation, fungus infection, vertebrate grazing and bacterial infection are some of the more common factors that can change the NP composition of a plant (see Chapters 5 and 8). Consequently, any analysis that is undertaken of the NP composition of a plant really only applies to the conditions used and a number of studies of the composition of plants grown under a range of conditions, with and without infestations and infections, would be required to provide more meaningful conclusions.

Thus, the value of metabolomics would currently appear to be greater as a research tool than as a universal tool to help assess the risks presented to humans or other organisms through the widespread cultivation of a plant with a changed NP composition.

Conclusion

The Screening Hypothesis was based on the simple idea that potent, specific biological activity is a very rare property for a chemical to possess. The hypothesis predicted that evolution would have favoured plants and microbes that possessed metabolic traits that enhanced the production and retention of NP diversity. Most of the traits predicted 15 years ago have been found; hence, the model has, so far, had a reasonable predictive value. The hypothesis predicted that these same traits would make the manipulation of pathways leading to NPs unpredictable. However, even if the genetic manipulation of an organism does cause it to produce some unexpected new products, the Screening Hypothesis suggests that these new chemicals have a very low probability of harming most consumers. Even if the new chemicals do possess some biomolecular activity that would be potentially harmful to the consumers, all consumers of NPs will have evolved generic methods of keeping the concentration of all ingested NPs low. In humans, this generic protection against NP accumulation must protect us efficiently from the thousands of NPs that a human might encounter in a modern, very varied, and often highly spiced diet.

Thus the Screening Hypothesis predicts that the manipulation of the NP composition of plants will produce unknown outcomes but there is only a low probability of harm to human consumers. However, will the public be reassured by what in effect is a probability argument? I would suggest that there is a high probability that they will not.

What does this chapter tell us about the way science works?

When scientists first successfully inserted genes into microbes and plants, it was the potential to exploit this science commercially that was sold to, and excited, politicians and investors. The governments of nearly all developed countries made sure that funds flowed into this new, exciting area of science, often at the expense of other areas of biology. This flow of funds automatically resulted in a bonanza for those with a training in molecular biology, with every university or research institute hiring new staff to work on this new, well-funded topic. Coincidentally or not, the early days of genetically modifying organisms corresponded with the acceptance that academic work should enrich society (and possibly academics). This was the first period of biological research when wealth generation and pure research became intertwined. It was also a time when the public became increasingly sceptical of science, a scepticism fed by the media. However, few of the public realised that the science they heard about in the media was science as revealed by press release from interested parties. The major science journals, very profitable ventures in themselves, feed journalists with predigested stories every week, in order to boost their own prestige. The science journal editors make sure that they hit the right buttons with the popular media; therefore, simplified, gross generalisations soon enter any debate in the popular media. Companies, universities, funding agencies and NGOs all learned the PR tricks with varying degrees of skill. Nowhere was this distortion of a proper scientific debate more evident than in the GM controversy. Press releases from the advocates of GM and from their opponents about the safety of GM foods, or the potential environmental harm that might result from growing weed-free or pest-free crops, polarised the debate. The public began to question the impartiality of scientists. Even the UK government found it difficult to identify GM scientists in the United Kingdom who could appear before the media without being vulnerable to claims that that person had a vested interest in supporting GM—many government institutes and universities had accepted money from GM companies. Scientists were slow to realise that once their work had become associated with wealth creation, their motives would inevitably be questioned. The GM food saga revealed, once again, that the mechanisms that exist to discuss science policy, which were never perfect, were seriously deficient in the new era of the biased press release.

Notes

Chapter 1: What Are Natural Products?

1. Given that how economically important Natural Products (NPs) are to humans, some readers might be surprised to know that no complete inventory of the NPs of any species has ever been published. The genomes of several species that make NPs are known but the NPs that these organisms can make are unknown. As explained in Chapter 9, many enzymes involved in NP biosynthesis are multifunctional and some NPs can be made by more than one route; hence, to assign individual genes to specific steps in NP biosynthesis might be a challenge.
2. The author acknowledges that the historical perspective adopted is highly biased towards 'Western' writings but hopes that others will provide a more comprehensive and balanced analysis of the contributions of other cultures to the study of NPs. Several sources were used to inform the author, including Kornberg A. (1997). Centenary of the birth of modern biochemistry. *Trends Biochemical Science*, 22, 282–3; Leicester HM. (1974). *Development of biochemical concepts from ancient to modern times*. Harvard University Press, Cambridge, Mass; Needham J. (1970). *The chemistry of life: lectures on the history of biochemistry*. Cambridge University Press, Cambridge.
3. Valency is the potential of any element to form multiple links to other elements. For example, hydrogen has a valency of one, oxygen a valency of two and carbon a valency of four.
4. It is ironic that the belief in vitalism, once central to the scientific study of naturally occurring chemicals, lives on in the minds of some non-scientists. The term 'Organic' is used in some English-speaking countries to identify 'natural' or 'ecological' products, products that are untainted by the endeavours of chemists. Those who believe in unique properties of 'Organic' products are essentially followers of Berzelius's classification of chemistry into inorganic and organic and the belief that living organisms possess some unique properties (the vital force).
5. As explained in Chapter 2, it can be argued that the concept of the commercial 'brand' was first used successfully to sell NP-rich products, especially cigarettes and soft drinks.
6. The concept of a scientific 'fact' is a slippery one. 'Facts' are a human construct and they have a remarkable way of changing as humans learn, think and experiment. As very well illustrated in the wonderfully readable book, Bryson B. (2003). *A short history of nearly everything*. Doubleday, Canada, each generation thinks that they have discovered the 'true facts', yet later generations usually show that previous understanding was not perfect. Consequently, all scientists should always be willing to challenge orthodoxy.

Chapter 2: The Importance of NPs in Human Affairs

1. Turner J. (2004). *Spice—the history of a temptation*. HarperCollins, London.
2. Bradford E. (2000). *Mediterranean. Portrait of a sea*. Penguin, London.
3. Armstrong K. (2001). *Islam: a short history*. Phoenix, London.
4. The Venetians did a remarkable job of exploiting crusaders. The Fourth Crusade, launched in 1199, commissioned the Venetians to transport the armies to Alexandria but the Venetians managed to get the crusaders first to attack the Christians ruling the growing Adriatic port of Zara, a commercial competitor of Venice. Then, the Venetians decided that the crusaders might

usefully invade Byzantium, rather than Egypt, to increase the Venetian control of that important trading region. So, instead of fighting Muslims, the crusaders fought the Christians and sacked the city of Constantinople. The Venetians even got paid for their devious work as well as gaining control of more strategic ports. However, the weakening of the power on Byzantium was to have terrible consequences within a short period.

5. Milton G. (1999). *Nathanial's nutmeg—how one man's courage changed the course of history.* Sceptre, London.
6. Allen SL. (2001). *The devil's cup—coffee, the driving force in history.* Canongate, Edinburgh.
7. Ellis M. (2005). *The coffee house: a cultural history.* Phoenix, London.
8. Hobhouse H. (1999). *Seeds of Change: six plants that transformed mankind.* Papermac, London.
9. Davenport-Hines R. (2001). *The pursuit of oblivion—a global history of narcotics 1500–2000.* Weidenfeld and Nicolson, London.
10. Perkin befriended and employed the young Jewish chemist Chaim Weizmann who used his knowledge of coal tar chemistry to devise a method of making acetone, which was necessary to make explosives during the First World War. Weizmann became director of chemical research at the British Admiralty when Arthur Balfour was the First Lord (1915–1916). Balfour was much impressed by Weizmann's contribution to the war effort and when Balfour moved on to be British Foreign Secretary he made the famous Balfour Declaration which promised the Jews a national home in Palestine. Weizmann became the first President of Israel when it was finally established after the Second World War. Thus, the strange links between the NP quinine, a failed synthesis, an accidental discovery of great importance and the eventual founding of a state which had been central to world tension throughout the past half century.
11. The search for a synthetic source of quinine led to the growth of the synthetic dye industry and ultimately to the development of the pharmaceutical industry (see also Chapter 2). The growth, and economic importance of the German dye industry in particular, was a major stimulus to the blossoming subject of synthetic chemistry in the late nineteenth century, and a very large number of synthetic dyes of all shades and hues were developed. The chemical stability of such compounds was essential for their use and some were much more stable than natural vegetable dyes. Thus, when Paul Ehrlich was given the task by the German chemist Hoffman (who had also taught Perkins while employed for a period in London) of seeking to establish the path of infection of malaria, Ehrlich found that methylene blue allowed him to trace the protozoan in an infected sailor who seemed to recover. The concept of selective toxicity began to grow. Clearly, a parasitic cell that took up a dye to such an extent that it became visibly stained against a background of cells that were not stained suggested that the concentration achieved in cells by dyes that penetrated them varied depending on the cell type. Given that for centuries it had been appreciated that poisons only acted when they were administered above a certain dose, it was reasonable to think that it might be possible to give high (= poisonous) doses of a chemical to some cells while leaving other cells unharmed. Ehrlich pioneered this approach and in doing so laid the foundation for modern methods of drug discovery. However, a drug to treat malaria eluded him. In 1905, the first drug to treat sleeping sickness, trypan red, was reported. In 1910, he discovered Salvarsan as a treatment for syphilis. However, it took his successors in Germany to finally discover the effective antimalarial drug. Pamaquin was discovered in 1926, and the drug mepacrine was developed in the United States and in the United Kingdom a few years later. These drugs were widely used in the Second World War.
12. Hobhouse H. (2003). *Seeds of wealth—four plants that made men rich.* Pan Books, London.
13. Gately I. (2002). *La Diva Nicotina—the story of how tobacco seduced the world.*
14. Schivelbusch W. (1993). *Tastes of paradise—a social history of spices, stimulants and intoxicants.* Vintage, New York.

15. State tobacco monopolies are still common—China has the largest but the state is still involved in the industry in Japan, Korea, Egypt, Turkey, Austria and so forth.
16. One lover of snuffboxes was Napoleon, who it is reputed consumed 1 kg of snuff a week! Napoleon also improved France's finances by enforcing tobacco taxation and his armies found tobacco a useful appetite suppressant when food supplies were limited.
17. Devine TM. (1999). *The Scottish nation 1700–2000*. Penguin, London. Some of the wealth that flowed into Scotland from its illegal and legal trade in tobacco was invested in shipping (to carry the tobacco from, and to carry goods to, the United States). Consequently, ship building blossomed near Glasgow to support the rich new shipping lines. One could argue that the city of Glasgow, like Venice or Amsterdam, grew from NP wealth. The bankers who controlled these investments in the west of Scotland were usually based in Edinburgh and the wealth flowing into Edinburgh helped build the much admired New Town and no doubt stimulated the remarkable intellectual flowering that characterised the Scottish Enlightenment. A few Scots had also become heavily influential in the other NP trading organisations in the United Kingdom, having become especially important in the East India Company, leading to the Scots playing a disproportionate role in the development of the British Empire, which further helped the Scottish industries that sprang from the tobacco wealth.
18. Lee D. (2000). *Nature's palette—the science of plant colour*. University of Chicago Press, London. Visible light is a term used to describe the wavelengths of light that human eyes can sense (roughly from the deep blue around 400 nm to the deep crimson of around 700 nm). Non-human species may have a different spectrum of light sensing.

Chapter 3: The Main Classes of NPs

1. As in the case of any spatial map, the visualisation is only an aid to navigation and need not represent reality. One of the most famous and influential maps, Harry Beck's London Underground, provides an excellent conceptual model by abandoning physical reality. It is interesting that some 'metabolic pathway' maps are drawn in a way that is remarkably like Beck's elegant style.
2. These chemicals made by the majority of organisms, hence considered essential for life, were categorised by Kössel as *primary metabolism* but as explained in Chapter 9. Kössel's classification was seriously flawed.
3. Understandably, the metabolic maps introduced in elementary biochemistry teaching focus on the few hundred 'primary metabolites' and rarely note the fact that these maps represents may be only 1% of the total metabolism found in organisms. It is as if a map of central London is used to teach UK geography.
4. Classification—convenient or meaningful? In order to bring a bit of order to any complex subject, categorisation is frequently applied. This is useful as long as one recognises that the categories might simply be chosen for convenience and the allocation of any item to a subcategory might be only one way of looking at the item. Consider, for example, the classification of books or wines. A detective novel might also be a historical novel or could even be science fiction. This problem was so serious that every book sold now has an agreed classification printed in it. A wine might be red, white or rosé but it might also be sweet or dry, an Australian wine or a German wine, in a clear bottle or a coloured bottle. Clearly, in both examples, there is no adequate way of categorising any individual item because each item has many properties—using only one property for categorisation misses some other important properties. The same goes for categorising chemicals. Any molecular structure will give that molecule several properties and one cannot safely use any single categorisation to judge the value or otherwise

of a molecule. Unfortunately, when one learns chemistry it is often one of the property classes that is given great prominence—alcohols, acids, esters, aldehydes and so forth. Not surprising because they are the key properties that a chemist finds useful to understand the properties that interest them. Likewise, it is often very difficult for biologists to think about chemicals without having a strong bias towards the biological properties of a molecule—insecticides, herbicides and so forth. Once one accepts that each molecule has several 'physical' properties (chemical stability, polarity, spectral features, etc.) and could even have more than one 'biological' property (ability to inhibit process A, ability to inhibit process B, susceptibility to microbial degradation, etc.), it is clear that only by judging the many properties of a molecule can one come to a conclusion about its fitness for use for one purpose. Ideally, you want each property of the molecule to match the 'property needs list' of the use.

5. Physicochemical is a term used to describe the properties of a substance that relate to the physical and/or chemical characteristics. Thus, a coloured substance absorbs certain wavelengths of visible light due to the way in which certain types of chemical bonds within the molecule interact. Likewise, a waxy or fatty substance possesses the property that we call fatty or waxy because of the way in which a combination of hydrocarbon bonds give the structure a non-polar characteristic. See Chapter 9 for a more detailed discussion.
6. Isoprene is made by animals and plants and is the most common volatile hydrocarbon found in the human body—the average person makes nearly 20 mg daily. However, this amount of isoprene can be released by 1 m^2 of plant biomass per hour under hot sunny conditions. The total release of isoprene per year by plants, 400–600 Tg of carbon, is a very significant contribution to global atmospheric chemistry.
7. Lange MB, Rujan T, Martin W, Croteau R. (2000). Isoprenoid biosynthesis: the evolution of two ancient and distinct pathways across genomes. *Proceedings of the National Academy of Sciences*, 97, 13172–7. See also Christianson DW. (2007). Roots of biosynthetic diversity. *Science*, 316, 60–1.
8. The fact that animals possess an 'NP pathway' is yet again evidence that the problem of classifying naturally produced chemicals as NPs is not inconsiderable. Chapter 9 begins to address the problem.
9. Ridley CP, Lee HY, Khosla C. (2008). Evolution of polyketide synthases in bacteria. *Proceedings of the National Academy of Sciences*, 105, 4595–600.
10. Jenke-Kodama H, Sandmann A, Müller R, Dittmann E. (2005). Evolutionary implications of bacterial polyketide synthases. *Molecular Biology and Evolution*, 22, 2027–39.
11. McDaniel R, Ebert-Khosla S, Hopwood DA, Khosla C. (1995). Rational design of aromatic polyketide natural products by recombinant assembly of enzymatic subunits. *Nature*, 375, 549–54.
12. Joseph PN, Austin MB, Bomati EK. (2005). Structure–function relationships in plant phenylpropanoid biosynthesis. *Current Opinion in Plant Biology*, 8, 249–53.
13. Juvvadi PR, Seshime Y, Kitamoto K. (2005). Genomics reveals traces of fungal phenylpropanoid-flavonoid metabolic pathway in the filamentous fungus *Aspergillus oryzae*. *Journal of Microbiology*, 43, 475–86.
14. Chen B. (2000). In FJ Leeper, JC Vederas (Eds.), *Biosynthesis: aromatic polyketides, isoprenoids, alkaloids* (Topics in Current Chemistry 209). Springer, Berlin/London, 1–52
15. The estimates of the number of alkaloids, like estimates of the total number of NPs or the number of any other groups, vary somewhat between different authorities. Even when commercially important plants have been thoroughly studied with respect to their NP composition, the analysis is never exhaustive simply because it gets harder, so more expensive, to find and characterise the substances that are present at low concentrations or those chemicals that are unstable.

16. Roberts MF, Wink M. (Eds.) (1998). *Alkaloids—biochemistry, ecology and medicinal applications.* Plenum Press, New York and London.
17. There have been reports of alkaloids in bacteria and it is possible that some alkaloids in marine invertebrates are actually made by bacteria that live on the surface of invertebrate (see Walls JT, Blackman AJ, Ritz AD. (1995). Localisation of the amathamide alkaloids in surface bacteria of *Amathia wilsoni* Kirkpatrick. *Hydrobiologia*, 297, 163–72). It would appear that the search for alkaloids in microbes has been less extensive than in higher plants. It is reported that alkaloids are much less common in lower plants and gymnosperms than in angiosperms.
18. McKey D. (1974). Adaptive patterns in alkaloid physiology. *American Naturalist*, 108, 305–20.
19. Textor S, de Kraker J-W, Huase B, Gershenzon J, Tokuhisa JG. (2007). MAM3 catalyzes the formation of all aliphatic glucosinolate chain lengths in *Arabidopsis. Plant Physiology*, 144, 67–71.
20. Wleker M, von Dohren H. (2006). Cyanobacterial peptides—nature's own combinatorial biosynthesis. *FEMS Microbiology Reviews*, 30, 530–63.

Chapter 4: Are NPs Different from Synthetic Chemicals?

1. Cordell GA. (2000). Biodiversity and drug discovery—a symbiotic relationship. *Phytochemistry*, 55, 463–80.
2. The term 'biological activity' is a commonly used term meaning that a chemical substance has the potential to bring about some form of effect on some organism. In Chapter 5, it is argued that the term is very ambiguous and best used with great care. Once one begins to probe how chemicals bring about their effects on organisms, the term 'biomolecular activity' (introduced in Chapter 5) is much less ambiguous, hence preferable.
3. Schneider G, Baringhaus KH. (2008). *Molecular design: concepts and applications.* Wiley-VCH, Weinheim.
4. Man-Ling Lee, Schneider G. (2001). Scaffold architecture and pharmacophoric properties of natural products and trade drugs: application in the design of natural product-based combinatorial libraries. *Journal of Combinatorial Chemistry*, 3, 284–9.
5. Lax L. (2004). *The mould in Dr Florey's coat.* Abacus, London.

Chapter 5: Why Do Organisms Make NPs?

1. The scientific publishing industry encouraged this fragmentation because it profited very greatly from it. In the last quarter of the twentieth century scientific publishing became extremely profitable, not only for the big commercial publishers (the notorious rogue Robert Maxwell was one of the first to recognise the fact that scientists were easily exploited) but also for the 'learned societies'. Indeed some learned societies became very wealthy and powerful by exploiting public bodies such as university libraries and government research institutions! Not only did the library subscription rates increase annually at levels well above inflation but publishers looked for 'gaps in the market' where they could try to launch 'must have' journals that only needed to cater for a small number to become very profitable. By getting an emerging scientific leader to edit a new journal catering to an increasingly topical area, university libraries could be milked a little more. Very healthy profits could be made even with a circulation of only several hundred (profits of $100–200 per subscription were possible on some specialist journals). The author speaks here as one who once took an active part in this trade.

2. In the author's opinion, the most accessible account of the way in which scientific thinking in all disciplines was developing in the eighteenth century and nineteenth century can be found in the following book: Bryson B. (2004). *A short history of nearly everything.* Black Swan, London. Bryson makes scientists seem very human and describes how science progresses despite the many obstacles often placed in its path by its practitioners.
3. Try making a list of all the different foods you have eaten in the past 24 hours and think of the challenges that you have given your biochemistry. Hundreds of NPs must have entered your system, most of which will be chemically unknown and of unknown toxicological consequence. Clearly, human experience has eliminated from the diet those foods that overburden the human body with harmful NPs. However, it is worth knowing that some plants (e.g., garlic) contains some very toxic NPs, chemicals that were they made synthetically, would not be allowed to be added to food.
4. This problem with the use of the term 'waste' is especially topical because there is a sudden enthusiasm by some people to gather up this 'waste' to be processed into energy for human use. Humans can burn cow dung or allow the dung beetle to use it. But if humans take it all, predictably the dung beetle population will fall. The same applies to 'waste straw' in fields (currently used by soil organisms), 'waste timber' in forestry (currently sustaining a range of species in forests) or any other form of 'waste'. This is obvious yet even some respectable scientists choose to overlook the problem because money is available to divert 'waste' into human use.
5. Zähner H, Ankle H, Ankle T. (1983). Evolution and secondary pathways. In JW Benett, A Ciegler (Eds.), *Secondary metabolism and differentiation in fungi* (pp. 153–175). Marcel Bekker, New York.
6. Davies J. (1990). What are antibiotics? Archaic functions for modern activities, *Molecular Microbiology* 4, 1227–32.
7. Fraenkel G. (1959). The raison d'etre of secondary plant substances. *Science*, 125, 1466–70.
8. Müller had shown in a brilliant, simple experiment that drops of water in contact with infected pea pods were more inhibitory when added to a fungal culture than drops that had been in contact with uninfected pea pods. The antifungal chemical produced by peas was isolated, characterised, and given the name pisatin. The work was highly influential and became a model for many other studies of the phytoalexins (antifungal chemicals that were made by plants in response to fungal attack). See Chapter 8.
9. Insects that were resistant to the concentrations of hydrogen cyanide, which was used to treat fruit trees in California to kill overwintering pests appeared within years of the adoption of the technique in the first decade of the twentieth century. The tally of pests and diseases resistant to specific control agents rise annually. About 500 insect pest species have evolved some resistance to some insecticides. Strategies to retard the evolution of resistant fungal strains are now a normal part of any programme to develop a new fungicide simply because a fungicide which fails to provide adequate control on a disease after only a few seasons use is commercially doomed.
10. Ehrlich PR, Raven PH. (1964). Butterflies and plants: a study in co-evolution. *Evolution*, 18, 586–608.
11. Demain AL. (1995). Why do microorganisms produce antimicrobials? In PA Hunter, GK Darby, NJ Russell (Eds.), *Fifty years of antimicrobials: past perspectives and future trends* (pp. 205–28). Symposium of the Society for General Microbiology. Cambridge University Press, Cambridge.
12. Bennett JW. (1995). From molecular genetics and secondary metabolism to molecular metabolites and secondary genetics. *Canadian Journal of Botany*, 73, S917–S924.
13. Jones CG, Firn RD. (1991). On the evolution of plant secondary chemical diversity. *The Philosophical Transactions of the Royal Society*, 333, 273–80. Despite the overwhelming evidence that most NPs possess no potent, specific biological activity, NP researchers often imply the opposite by using generalisations that imply the opposite. For example, 'leaf alkaloids

protect induced foliage from attack' or 'alkaloids protect the organism against predators, competitors or pathogens' or 'naturally occurring alkaloids protect the wild potato plant from bacteria, fungi and insects' when the reality is that *some* alkaloids might do so but most do not. One can find such misleading generalisations applied to all classes of NPs, repeated like creeds.

14. Firn RD, Jones CG. (1996). An explanation of secondary product 'redundancy'. In JT Romeo, JA Saunders, P Barbosa (Eds.), *Phytochemical diversity and redundancy in ecological interactions* (pp. 295–312). Plenum Press, New York.
15. Unfortunately, unknown to the authors of the Screening Hypothesis to explain NP diversity, the term Screening Hypothesis was used by educationalists to postulate that education acted as a filter to screen those that can be trained. That version of the terminology currently wins a Google search.
16. The term biological activity is a term that has little more scientific value than a term like big. The adjective big is only useful in scientific debate when it is given a reference point. If an organism is identified as being big, is it big relative to others of the same species, big relative to all species or simply big relative to humans? Big is a useful term if it is given a context and so is the term biological activity. As of late 2008, a Google search for 'biological activity' finds over 5.5 million hits so the term may be vague but it is well used.
17. There are some natural insecticides known but only the pyrethrums show useful selectivity. The old natural insecticides, such as derris, rotenone and nicotine, are now rarely used commercially because they are considered more environmentally damaging than synthetic alternatives. The author knows of no commercially significant NP fungicides.
18. Firn RD, Jones CG. (2003). Natural products—a simple model to explain chemical diversity. *Natural Product Reports*, 20, 382–91.
19. The choice of the word 'bind' to describe the association of a small molecule with a larger protein molecule is an unfortunate one because in English the word 'bind' can mean a physical link (e.g., to tie together physically with string), but it can also mean simply to unite two entities in some way without any physical link (e.g., 'people are bound by family ties'). The great majority of interactions between small molecules and proteins do not involve any physical joining, they are a much weaker interaction. There are a few examples of small molecules acting irreversibly by chemically linking to their receptor (see 32) but such interactions are very rare in nature.
20. Lenser T, Gee DR. (2005). *Modelling the evolution of secondary metabolic pathways.* University of York, MPhil Project Report (*Abstract*). Plants and microbes invest heavily in producing chemicals termed Natural Products. These chemicals are produced in secondary metabolic pathways. In this report, we develop a model for the evolution of secondary pathways, and investigate what factors are important in allowing these pathways to arise and persist. The results imply that certain mutation rates are important in generating chemical diversity, and we give conditions on these for optimal fitness in a population. We also find that the rate of competitive evolution and the chances that new compounds have to be beneficial or harmful are important factors.
21. Firn RD, Jones CG. (2009). A Darwinian view of metabolism: molecular properties determine fitness. *Journal of Experimental Botany*, 60(3), 719–26.
22. Firn RD, Jones CG. (2000). The evolution of secondary metabolism—a unifying model. *Molecular Microbiology*, 37, 989–94.
23. Schwab W. (2003). Metabolome diversity: too few genes, too many metabolites? *Phytochemistry*, 62, 837–49.
24. Croteau R, Karp F, Wagschal KC, Satterwhite DM, Hyatt DC, Skotland CB. (1991). Biochemical characterisation of a spearmint mutant that resembles peppermint in monoterpene content. *Plant Physiology*, 96, 744–52.

25. Steele CL, Crock J, Bohlmann J, Croteau R. (1998). Sesquiterpene synthases from grand fir (*Abies grandis*). Comparison of constitutive and wound-induced activities, and cDNA isolation, characterization, and bacterial expression of delta-selinene synthase and gamma-humulene synthase. *Journal of Biological Chemistry*, 273, 2078.
26. Berenbaum MR, Zangerl AR. (1996). Phytochemical diversity: adaptation or random variation. In JT Romeo, JA Saunders, P Barbosa (Eds.), *Phytochemical diversity and redundancy in ecological interactions* (pp. 1–24). Plenum Press, New York.
27. There was a definite reluctance of many working on NPs to admit that studies on the biological activity of synthetic chemicals could provide any useful information about the biological activity of NPs as a group. The fragmentation of the subject again allowed a group to ignore a huge amount of data simply because it did not fit their preconceived ideas. Partly this was because there was a lingering idea that NPs were somehow different from synthetic chemicals (see Chapter 4).
28. Possibly, the most extensive screening of NPs ever undertaken was conducted by the US National Cancer Institute, starting in 1960. Over two decades, 114,000 extracts from 35,000 plant samples (from over 12,000 species) were screened but less than 1% showed selective anticancer potential. One assumes each sample must have contained tens or hundreds of NPs so the hit rate was really much lower. However, the bioassays used have a very questionable relationship to any functional significance of endogenous NPs because plants do not form cancers in the manner that animals do (see Chapter 8).
29. Suppose when supplied at 1 micromolar, chemical A alone causes a 25% inhibition of a process and chemical B alone causes a 25% inhibition. But if when both chemicals were given at 1 micromolar in combination and the resulting inhibition was 50%, the effects of A and B would be said to be additive but if the inhibition was >50% then the effect would be said to be synergistic.
30. The view that the biological testing of chemicals in isolation from other chemicals may underestimate the potential of that chemical to cause harm has been advanced by many worried about the toxicity of synthetic chemicals (see also Chapter 6). For example, some environmentalists have argued that the toxicity testing, conducted as part of the accreditation process needed before any pesticide is sold, is fundamentally flawed because such tests do not take into account the possible interactions that could occur when multiple pesticides are used (and some crops are indeed exposed to several herbicides, fungicides and insecticides during a growing season). Likewise, the evaluation of the safety and efficacy of drugs has been questioned because each drug is studied largely in isolation from other chemicals to which the patient may be exposed. These criticisms have increasingly been taken seriously by regulatory bodies and worries have been expressed in the European Union that while a great deal is known about the toxicity of chemicals when they are used alone to treat organisms, insufficient information is known about the effects of exposure to multiple chemicals. Given that there maybe 80,000 synthetic chemicals released into the environment, the scale of the problem becomes apparent. The EU REACH regulations seek to address this issue with a further, very expensive evaluation of the toxicity of many chemicals currently in use. As discussed in the main text, organisms that make or are exposed to NPs will inevitably have been exposed to very complex mixtures of NPs.
31. Berenbaum MR, Zangerl AR. (1986). Variation in seed furanocoumarin content within the wild parsnip *Pastinaca sativa*. *Phytochemistry*, 25, 659–61.
32. Baker BR. (1967). *Design of active-site-directed irreversible enzyme inhibitors*. John Wiley, New York. Baker introduced the term 'active site irreversible inhibitors' to describe ligands that were drawn closely into the active site of a target protein by a high-binding affinity but once the chemical was intimately associated with the protein, a chemically reactive group on the ligand

would react with an amino acid in the protein to form a covalent bond. The protein would now have a permanently changed structure, with some properties destroyed, due to the linked chemical. 'Affinity reagents' and 'photoaffinity reagents' were synthetic chemicals designed to react covalently in this way and they were very useful tools for biochemists to probe protein structures. However, one problem with this technique was that all too often the 'affinity reagents' or 'photoaffinity reagents' did not only react exclusively with the target protein but also reacted with many non-target proteins. This lack of speficity explains why such reagents have no clinical or agrochemical value. This lack of specificity would also explain why evolution has largely eliminated chemically reactive substances during the evolution of NPs (one exception being the furanocoumarins discussed in the text).

33. Some non-scientists, and sadly too many scientists, seem to think that science advances only when new techniques or equipment allow novel experiments to be conducted. The media tends to encourage this view because expensive kit always looks impressive (the finest recent example is the Giant Hadron Collider at CERN). However, sometimes it is ideas that are needed, and ideas are hard to capture visually and often difficult to explain in simple terms. Worse still, ideas cost nothing; hence in a society where money is everything, ideas seem unimpressive. This attitude has even infected science management in many countries where schemes have been set up to identify the 'best' science by judging the cash inputs rather than the long-term scientific outputs. It may interest some readers to know that the author of this book has never received any financial support for any of his work on NPs.

Chapter 6: NPs, Chemicals and the Environment

1. In 1884, the Danish microbiologist Hans Christian Gram showed that the sequential treatment of bacteria, first with the purple dye *crystal violet*, then with iodine and at last a washing solution of alcohol, resulted in some species becoming pigmented. It was subsequently shown that the dye and iodine alone can diffuse into any cell but once inside the cell, the iodine interacts with the purple dye to form a chemical so large that it gets trapped in certain species of microbe, species which have a type of cell wall around the cell that makes it hard for large molecules to penetrate.
2. The word 'screening' means to use a method to isolate one item from many, on the basis of some property that the desired item might possess that the majority of items do not. The use of a *screen* or sieve to separate items on the basis of size was common in agriculture (seed cleaning), food production (flour) and manufacturing (sand, gravel, pigments, etc.). Biologists use the term to mean testing a large number of samples for one property.
3. Examples of selective toxicity were already known with the use of the poisonous metallic salts used to control fungi and weeds on crop plants. For example, Millardet, in 1895, had noticed that a mixture of copper sulphate and lime, applied to grapes alongside roads near Bordeaux to deter those on foot from stealing the ripe grapes, protected the grapes from mildew infection (the copper had an adverse effect on the mildew fungus without harming the vines). Soon 'Bordeaux Mixture' was being sold to farmers to protect potatoes from the fungus that causes blight (*Phytophora infestans*), a use which continues. The ability of sulphuric acid to kill some weeds in fields of grain, without harming the grain crop itself, is another example of differential uptake, where the waxy upright leaves of the wheat or barley shed the spray but the horizontal broad leaves of the weeds retain it.
4. Some have since argued that most of the wildlife changes detected in the 1950s, were mainly a consequence of the increasing adoption of the intensive, highly mechanised agriculture that characterised the post-Second World War food production. In northern Europe, for example,

the centuries old patterns of agriculture changed due to mechanisation and the introduction of new crops, especially winter sown cereals to replace spring sown crops.

5. It is now known that there are many ways, other than direct toxicity, in which the use of a pesticide can adversely affect the population of a non-target organism. Suppose an insecticide is sprayed on a field to control a particular insect pest. The chances of any adult bird being directly sprayed are very low. More vulnerable are chicks, especially if they are fed insects that are dying from the insecticide exposure. However, maybe the biggest problem will be the starvation of the insect-eating chicks, if the insecticide works well. Likewise, a herbicide that is not directly toxic to birds can remove the food source (weed seeds) from a bird population.
6. Not all of the evidence was sound because the early analytical methods for DDT using gas chromatography could confuse PCBs (polychlorinated biphenols) and DDT breakdown products. However, these analyses alerted scientist and the public to the widespread contamination of the environment with PCBs (widely used in electrical transformers and other industrial equipments at that time but subsequently banned), which became a new concern.
7. Green MB, Hartley GS, West TF. (1987). *Chemicals for crop protection and pest control* (3rd edn.). Pergamon Press, Oxford.
8. Controversially, the politicians had responded to lobbying by the agrochemical companies not to make the new, more extensive safety testing retrospective immediately. Clearly, the removal of all products from the market for a new prolonged period of reassessment would harm the agricultural economy. Furthermore, it was argued that many products had been in widespread use for decades, with no evidence of harm, so such products had effectively been safety tested in the real world. However, this compromise did allow some rather doubtful chemicals to remain in use before being slowly phased out without ever being fully reassessed.
9. Many animal husbandry problems are caused by infectious agents. Fungal diseases, such as foot rot in sheep, were treated by chemicals (copper sulphate) that had also found a use in controlling potato blight. Insects causing damage to hides in cattle or infestation of sheep were controlled by insecticides related to insecticides in crop use—as is the treatment of head lice in humans. The treatment of helminths (worms) in pigs is not dissimilar to the treatment of worms in humans. There are really no conceptual barriers to the exploitation of selective toxicity in any area of biology. Hence, the view that the agrochemical and pharmaceutical industries can be distinguished in terms of the scientific principles is wrong and unhelpful.
10. The inability of the public and of politicians to make rational judgements of risk is well documented. The key elements of an individual, in judging a risk are the degree of control the person has over the hazard, the dread of the outcome and the drama that might ensue from an accident. Thus, driving a car that carries a very much greater risk than eating food-containing additives and pesticide residues is judged safer by most consumers. This is because the driver is in control of the car but has no control of what is in their food. The car driver also tends to think of physical injury as the most likely outcome of a car accident whereas they often associate chemicals with the more dreadful cancer.
11. The EU Drinking Water Directive suffered from another real problem—there are multiple sources of pesticides in an individual's diet; hence, the elimination of pesticides from drinking water is hardly likely to help much if the food one consumes contains much more significant doses of the same pesticides that were expensively reduced in the water. The old toxicology rules ignored by politicians for drinking water are still applied to foods and allow consumers to eat significant amounts of many pesticides!
12. That number is so big that it means little to most people. Try thinking of the total mass of all humans, cows, pigs and sheep added together and you are getting there. It is possible that this is an underestimate because it is based on plant productivity and as plant material is degraded by microbes more NPs can be made.

13. Many human populations have only encountered some mixtures of NPs relatively recently—the first Europeans to encounter chilli peppers, many beans, pineapples, bananas, tobacco and so forth did so only a few hundred years ago. The solanine alkaloids (and other NPs) in potato or tomato were unknown to Asians, Africans and Europeans until very recently, yet these populations seemed untroubled by these novel chemicals. Tomato fruits were initially considered to be poisonous when introduced to Europe and it was only in the nineteenth century that they became widely adopted as a food plant.
14. Organisms that have a very restricted diet can be expected to have evolved more specific mechanisms to target chemicals in the diet that are especially toxic substances. Thus, an insect living on one plant might protect itself against one toxin by specific biochemical traits—see Chapter 9.
15. However, resistance development just after the patent life expires helps to keep cheap generics out of the market, leaving the newer, patented, more expensive products to find a market.
16. An analogy would be that a method of detecting people, which only found those with red hair, would give a very misleading impression of the population of London.
17. The concentration of attention on synthetic chemicals is partly due to the need to fulfil regulatory requirements to study the fate of pesticides and other major industrial chemicals and partly due to the fact that the presence of NPs in the soil or in water has not been generally regarded as of importance or significance. It is also important to recognise that some degradation of NPs will occur through chemical degradation such as photo-oxidation.
18. For most of the twentieth century, it was assumed that all microbial organisms could be grown in culture if only the appropriate growth medium was found. However, when estimates are made, using molecular biological techniques, of the total number of microbial species in soil or water samples, it has been estimated that typically only 10% of the species present have ever been cultured.
19. A stimulating discussion of the way in which scientific personalities and vested interests can sway government thinking can be found in the following book: Taubes G. (2007). *The diet delusion*. Alfred A Knopf, New York.

Chapter 7: Natural Products and the Pharmaceutical Industry

1. Excellent popular accounts of the European nations', Japan's and to a lesser extent the United States of America's scramble for colonies is given by Eric Hobsbawn's 'The Age of Empire' or Thomas Packenham's 'The Scramble for Africa'. It is sometimes overlooked that it was only in the latter half of the nineteenth century that national governments became directly involved in colonisation. In the previous three centuries, it was largely the commercial exploitation of NPs that motivated a few Europeans to seek to control events in far off lands (see Chapter 2).
2. Hoffman, while employed for a period in London, had taught Perkins, the founder of the UK dye industry. Perkins was actually trying to make the NP quinine when he made the dye mauve!
3. The great Paracelsus (Phillip von Hohenheim, 1493–541) is credited with the first clear statement that all chemicals are poisons but every chemical has a non-toxic dose—'it is the dose that maketh the poison'. Most members of society, many politicians (see Chapter 6) and all parts of the mass media fail to appreciate the importance of this fact, a fact which was later formalised in the Law of Mass Action (see Chapter 5). In the United Kingdom, every year, a government agency publishes a survey of the presence of pesticide residues in food. One has yet to see the following headline to the inevitable story 'No toxic levels of pesticides found in food'.
4. Screening is a word borrowed from other human activities where a method is devised to efficiently separate wanted items from unwanted items. The sieve or screen can separate the grain

from the straw or the stones from the soil; in both cases, the grid in the sieve or screen selects on the basis of size. Gold prospectors use a pan to 'screen' their material using density as the criteria for selection—prospecting for gold is very similar to prospecting for pharmaceutical drugs in that the majority of material handled is valueless but the rare nugget can be worth a fortune or it might be 'fools gold'.

5. Until that time the main treatment for syphilis had been a prolonged treatment with the highly toxic and cumulative heavy metal mercury—'one night with Venus and a lifetime with Mercury'.
6. The term 'pharmaceutical company' has gradually been replaced by the term 'pharma' in financial and trade circles.
7. Pearce DW, Puroshothamon S. (1995). The economic value of plant-based pharmaceuticals. In T Swanson (Ed.), *Intellectual property rights and biodiversity conservation* (pp. 19–44). Cambridge University Press, Cambridge.
8. Lax E. (2005). *The mould in Dr Florey's coat—the remarkable true story of the discovery of penicillin*. Abacus, London.
9. This discovery provoked a large programme of the screening of sulphanilamide analogues and this programme provides further evidence that the probability that any substance possessing potent biomolecular activity is low. Around 5488 derivatives of sulphanilamide had been investigated by 1945, yet none could compete with penicillin.
10. As in the case of the discovery of penicillin, where some believe that the Nobel Committee did not give sufficient credit to the contribution of Norman Heatley, so Waksman has also been accused of giving insufficient credit to his graduate student Albert Schatz. Waksman's screening programme was initially undertaken by graduate students among whom was Albert, who started work in June 1942. Schatz was drafted into the army in November 1942 and was posted as a laboratory technician to a Medical Detachment of the Air Corps in Florida. He began a search for antibiotics that would be useful against bacterial diseases that were not susceptible to penicillin. However, he was discharged due to ill health and returned as a research assistant to Waksman but continuing the work he had begun in the army. Waksman was taking an interest in *Mycobacterium tuberculosis* and because the bacterium was so virulent, he found Schatz an isolated laboratory in the basement where he was encouraged to screen for antibiotics that would control the pathogen. It was Schatz who was fortunate in isolating an actinomycete, *Actinomyces griseus* (since renamed *Streptomyces griseus*) which produced a substance that was effective against a wide range of pathogens, including *M. tuberculosis*. Schatz received his PhD for his discovery of streptomycin and was the first author of the paper (by Schatz, Bugie and Waksman) announcing the discovery in 1944. (In a sworn affidavit, Bugie later credited the discovery to Schatz and Waksman and suggested that her contribution was minor. However, when the controversial royalty settlement was agreed, Bugie received a 0.2% share. Bugie later in life told her daughter 'They approached me privately and said, some day you will get married and have a family and its not important that your name be on the patent'—information from Drs. Milton Wainwright and Ross M Tucker.) Before Schatz left Rutgers, a patent was awarded in 1948 to Waksman and Schatz for streptomycin and a gentlemen's agreement was made that neither individual would profit from this discovery instead all royalties would go to the Rutgers Research Foundation. However, in 1949, problems began and the amicable relationship between the two took a turn for the worst. Schatz, now working at Hopkins Marine Station in California, received documents from Waksman asking Schatz to sign away his rights to credit and any royalties from streptomycin. Schatz learnt that Waksman had negotiated a contract with Rutgers to receive 20% of the royalties from streptomycin which by that time amounted to $350,000. Eventually, Schatz sued Waksman and Rutgers to the great embarrassment of all. However, rather than going to trial, a deal was reached between the parties where Schatz was

to be given credit as the co-discoverer of streptomycin as well as 3% of the royalties. Waksman received 10%, while Rutgers received the lion's share of 80%. The remaining 7% was distributed to all of the students and researchers who participated in the discovery. However, poor Schatz had further reason to feel aggrieved. His lawyer took 40% of his first $125,000 royalty and he was shunned by the scientific community because he had dared challenge authority. Despite the legal recognition of his contribution, Schatz still received no recognition from the Nobel Committee when Waksman was awarded the prize alone in 1952. It took decades before Schatz was given the credit he fought for and even Rutgers finally accepted his contribution by awarding him the Rutgers Medal as co-discover of streptomycin.

11. A patent in the name of CT Beer, JH Cutts (a doctoral student and co-worker) and RL Noble was administered by the University of Western Ontario in cooperation with the Eli Lilly Co. of Indianapolis. While Dr Noble has received broad recognition for this important work, Dr Beer's essential role in the vinblastine story has been largely overlooked.
12. Cordell GA. (2000). Biodiversity and drug discovery—a symbiotic relationship. *Phytochemistry*, 55, 463–480.
13. Balick MJ, Mendelsohn R. (1992). Assessing the economic value of traditional medicines from tropical rain forests. *Conservation Biology*, 6, 128–130. See also Svastad H, Dhillion SS. (2000). *Responding to bioprospecting—from biodiversity in the south to medicines in the north.* Spartacus Forlag, Oslo.
14. Dhillion SS, Amundsen C. (2000). Bioprospecting and the maintenance of biodiversity. In H Svastad, SS Dhillion (Eds.), *Responding to bioprospecting—from biodiversity in the south to medicines in the north* (pp. 103–32). Spartacus Forlag, Oslo.
15. When the much publicised case of the Merck investment in INBio is looked at more closely, it was clearly a much less enthusiastic endorsement by that company of the potential of bioprospecting. The Merck and Co. investment of $1 million in INBio was less than 0.1% of that company's Research and Development (R&D) budget in 1991. To put the figure of $1 million into perspective, in 1999 Merck & Co.'s sales were $13,693 million, their R&D budget was $1,821 million and one drug (Vioxx) had potential annual sales of $2,000 million. As the twenty-first century began, the US pharmaceutical industry was spending >$8,000 million on advertising and drug promotion alone. The investments by Merck & Co. in INBio and the Eli Lilly investment in Shaman Pharmaceuticals (which is no longer trading) were crumbs from the rich man's table not serious commitments to NP screening.
16. Fellows L, Scofield A. (1995). Chemical diversity in plants. In T Swanson (Ed.) *Intellectual property rights and biodiversity conservation* (pp. 19–44). Cambridge University Press, Cambridge.
17. Rausser GC, Small AA. (2000). Valuing research leads: bioprospecting and the conservation of genetic resources. *Journal of Political Economy*, 108, 173–206.
18. Firn RD. (2003). Bioprospecting—why is it so unrewarding? *Biodiversity and Conservation*, 12, 207–16. See also Macilwain C. (1998). When rhetoric hits reality in debate on bioprospecting. *Nature*, 392, 535–40.
19. Vioxx, a drug with $1.8 billion US sales in 2003, which was specifically designed to target only a cox-2, has been withdrawn by Merck and Co. after it was reported that some patients had died from side effects. A few individuals sued Merck & Co. for hundreds of millions of dollars—hundreds of times what Merck & Co. invested in INBio.
20. Firn RD, Jones CG. (1998). Avenues of discovery in bioprospecting. *Nature*, 393, 617.
21. Handelsman J, Rondon MR, Brady SF, Clardy J, Goodman RM. (1998). Molecular biological access to the chemistry of unknown soil microbes: a new frontier for NPs. *Chemistry and Biology*, 5, 245–9.

22. The generation of a toxic substance from an innocuous one is known. Such 'lethal synthesis' has been reported in both the pesticide and pharmaceutical industries.
23. Because so much attention was paid to the degradation of substances in mammalian livers, such metabolism was even taken as a reference point for some discussions of the ability of plants to degrade xenobiotics. Plants were even discussed in terms of their properties as 'green livers' (Sandermann H. (1994). Higher plant metabolism of xenobiotics: the 'green liver' concept. *Pharmacogenetics*, 4, 225–41). This seems unfortunate because plants had evolved the generic NP metabolic features long before livers evolved and it was animals evolving in response to plant and microbial NPs that drove the evolution of some of the properties of the liver.

Chapter 8: The Chemical Interactions between Organisms

1. Listen to a piece of your favourite music and think about what you are hearing. You will be very conscious of the complexity of the sounds you are hearing. Asked to listen to the drums, your brain will make you more conscious of the sounds characteristic of that instrument. Likewise, most people can selectively 'listen' to other aspects of the sounds they hear. Similarly, look up now and look around you. Your brain will process the visual images to detect pattern, colour, hue, movement and shape. If you are asked to describe what you see you will be able to give a rich description—you have the vocabulary and experience to share your sensory experiences with others. Now bite into an apple or sip your coffee or nibble your chocolate. Hundreds of NPs will enter your mouth but you will be unaware of most of the chemical complexity—your taste sense seems very crude compared to the visual and aural senses. Asked to describe the taste of coffee you will find your senses seem to let you down and your vocabulary lacking. Even if you ask a wine expert, with a much more discriminating palate, their vocabulary will be limited and still hinders communication. At best they might describe a wine as 'smokey', 'oaky', 'gooseberry' or 'citrus' relying on shared comparisons. But maybe other organisms sense chemicals with more sophistication? If dogs could speak, would they have hundreds of words to describe smells?
2. The development of highly selective 'antibiotics' for medical use (described in more detail in Chapter 7) illustrates the principles being discussed. It is clear with hindsight that when seeking highly selective antibacterial chemicals for use in humans a bias towards finding NPs that targeted proteins unique to culturable, pathogenic bacteria was inevitable—that gave the selectivity sought. Penicillin acts on an enzyme that is essential for making one type of bacterial cell wall and mammals do not make such cell walls, hence are unharmed by penicillin. Several other antibiotics (puromycin, actinomycin D, etc.) act on pathways used to make proteins or nucleic acids in bacteria, pathways that are not shared with higher organisms. So, selectivity is achieved by targeting proteins common in one's enemies but absent in one's friends. However, since the introduction of broad spectrum of antibacterial agents, it has become clear that humans have some very positive interactions with bacteria that are susceptible to antibiotics. For example, taking an antibiotic orally might help combat a bacterial infection that is causing grief to an individual but the broad spectrum antibiotic activity will kill much of the gut flora non-selectively with resulting negative health consequences for that person.
3. Because most readers will be unfamiliar with microbes, they may find it easier to think about these issues by using their knowledge of a higher organism such as a plant. A plant that can efficiently attract a pollinator, that can attract some insect species to attack herbivorous insects that feed on it or can make its seeds (gene containers, capable of preserving the genes in bad

times and moving the genes to new places) attractive to an organism that will disperse the genes, will be fitter than a plant that lacks these capacities.

4. A biofilm is a thin layer of microbes that adheres to a surface by means of secreted material and may be a mixed community of species. The total metabolic and physical capacity of the community provides an improved environment for the individuals to thrive (http://en.wikipedia.org/wiki/Biofilm).
5. The NP cycle is very hard to quantify. Many organisms can make and modify NPs; consequently, there are many routes around the cycle. However, the NP cycle could be between 1 and 10 billions of tonnes per annum because the annual photosynthetic fixation of carbon is about 100 billion tonnes and the primary producers can often use 1–5% of that carbon to make NPs. Furthermore, some microbes, also producers of NPs, live on the carbon fixed by the primary producers; hence, a further fraction of the annual carbon is converted to NPs. The author knows of no estimate of the flow of carbon into NPs in these organisms.
6. Clearly, single-celled organisms can move and show some simple behaviour so there is no absolute need to start this discussion at the point where multicellularity began but is convenient.
7. The first generation of commercially important insecticides, the organochlorines (DDT, Aldrin, Dieldrin, BHC), act on ion channels; hence, interfere with nervous transmission. The second generation organophosphates (Malathion, Bromophos, Diazinon, etc.) target acetylcholine esterase, the enzyme that degrades acetylcholine, an essential part of the neurotransmitter cycle. The third generation synthetic pyrethroids, developed to mimic the NP pyrethrum, also interfere with nervous transmission. Some of this bias was undoubtedly due to the first insecticide screening regimes being based on a rapid 'Knock down' effect which nearly inevitably selected for nerve action.
8. Cost–benefit analysis. An extensive literature exists on this topic, much of it related to models derived by economists (http://pondside.uchicago.edu/ecol-evol/faculty/bergelson_j.html).
9. The soil contains millions of seeds per hectare in 'the seed bank' and these individuals are easily overlooked as members of the plant community. The dormant seeds can survive for decades in the soil; hence, a population of annual plants does not have to be successful at reproducing every year in order to be sustainable. In contrast, a specialist insect herbivore of that annual species can rarely be sustained in a locality without a reliable source of food annually.
10. The perennial plant can reuse certain organs every year, hence need not build them anew. Furthermore, perennials can usually produce a full flush of foliage more quickly than an annual plant, hence maximises photosynthetic gain early in the season—in contrast to the rapid canopy cover of a tree with the rate at which an annual crop such as wheat, potato or maize reaches full coverage.
11. Agrawal AA, Karban R. (1998). Why induced defenses may be favored over constitutive strategies in plants. In R Tollrian, D Harvell (Eds.), *Ecology and evolution of inducible defenses* (pp. 45–61). Princeton University Press, Princeton, N.J.
12. But remember that in an earlier chapter it was noted that there is an analogy between the immune system and NP metabolism. Both are mechanisms evolved to generate chemical diversity to overcome the low probability of any one product having appropriate biomolecular activity.
13. Hammerschmidt R. (1999). Phytoalexins: what have we learned after 60 Years? *Annual Review of Phytopathology*, 37, 285–306.
14. De Vos M, Van Oosten VR, Van Poecke RMP, Van Pelt JA, Pozo MJ, Mueller MJ, Buchala AJ, Métraux JP, Van Loon LC, Dicke M. (2005). Signal signature and transcriptome changes of *Arabidopsis* during pathogen and insect attack. *Molecular Plant–Microbe Interactions*, 18, 923–37.

15. Whenever a screening programme finds 'a hit' (a chemical that shows some desired biomolecular properties), the company chemists make as many chemical variants as they can for several reasons. First, they might find a chemical variant with higher or more specific biomolecular activity. Second, they might find a variant that is cheaper to synthesise industrially while still retaining the desirable biomolecular activity. However, one of the main motivations is to ensure that any patent that is taken out to protect the investment on the chosen chemical is broad enough to ensure that competitor companies cannot enter the market simply by patenting and marketing a 'me-too' chemical. The extent of this screening of analogues of commercially successful products can be considerable. It is reported that in the mid-twentieth century, many agrochemical companies that existed at that time made and screened over 100,000 different organophosphorus chemicals as they searched for insecticides and over 100 different such insecticides eventually reached the market. The chemicals made specifically for the assessment of one purpose have sometimes been found having a quite different commercial use. The history of the ICI Plant Protection (a UK company that no longer exists) records that some of its most successful agrochemicals were found by accident. The company's first big success, the bipyridium herbicides, paraquat and diquat, were made by the ICI Dyestuff division; ICI's first successful systemic fungicide was made to screen as an insecticide and conversely the insecticide pirimicarb was made as part of a fungicide screening programme.
16. There will always be a cost to maintaining an inducibility system because it is likely that several proteins will be involved in the sensing of the threats and more proteins are likely to be involved in the control processes. Consequently, these maintenance costs could be saved but a loss of any part of this system might produce only a very minor cost saving.
17. Felton GW, Korth KL, Bi JL, Wesley SV, Huhman DV, Mathews MC, Murphy JB, Lamb C, Dixon RA. (1999). Inverse relationship between systemic resistance of plants to micro-organisms and to insect herbivory. *Current Biology*, 9, 317–320. See also Feys BJ, Parker JE. (2000). Interplay of signalling pathways in plant disease resistance. *Trends in Genetics*, 16, 449–55; Karban R, Baldwin IT. (1997). *Induced responses to herbivory*. University of Chicago Press; McDowell JM, Dangl JL. (2000). Signal transduction in the plant immune response. *Trends in Biochemical Science*, 25, 79–82; Karban R, Agrawal AA, Thaler TS, Adler LS. (1999). Induced plant responses and information content about risk of herbivory. *Trends in Ecology & Evolution*, 14, 443.

Chapter 9: The Evolution of Metabolism

1. Endersby J. (2007). *A guinea pig's history of biology*. Heinemann, London.
2. Firn RD, Jones CG. (2009). A Darwinian view of metabolism: molecular properties determine fitness. *Journal of Experimental Botany*, 60(3), 719–26.
3. Hartmann T. (2007). From waste products to ecochemicals: fifty years research of plant secondary metabolism. *Phytochemistry*, 68, 2831–46.
4. Hartmann T. (2008). The lost origin of chemical ecology in the late 19th century. *Proceedings of the National Academy of Sciences*, 105, 4541–546.
5. All readers will be familiar with the diversity of lipids in the form of plant 'oils' sold in supermarkets. Look at the range of plant oils on the shelves—soy, sunflower, safflower, maize (corn), rape (canola), sesame, olive, walnut, grapeseed, peanut, palm and so forth. Many plant oils are storage compounds, made by the plant to provide energy and carbon to sustain the growth of the germinating seed. The main chemicals in such oils are triglycerides—a glycerol backbone to which is joined three fatty acids. The type of fatty acid found in the storage lipid of any one species is its characteristic. Each species of plant has evolved its own lipid composition because it is the overall properties that the mixture of lipids gives to the plant that matters, not the way in

which the mixture is made up. The various species, like humans, can use most of these different oils interchangeably. Indeed, humans go one step further by blending whatever oil is cheapest at any one time to make and sell 'vegetable oil'.

6. Horowitz NH. (1945). On the evolution of biochemical syntheses. *Proceedings of the National Academy of Sciences*, 31, 153–7. See also Jensen RA. (1976). Enzyme recruitment in evolution of new function. *Annual Review of Microbiology*, 30, 409–425; Chapman DJ, Ragan MA. (1980). Evolution of biochemical pathways; evidence from comparative biochemistry. *Annual Review of Plant Physiology*, 31, 639–78.
7. Kaufman SA. (1993). *The origins of order—self organisation and selection in evolution.* Oxford University Press, New York.
8. Physicochemical means the combined chemical and physical properties of a substance. As defined in Wikipedia, 'All fats consist of fatty acids (chains of carbon and hydrogen atoms, with a carboxylic acid group at one end) bonded to a backbone structure, often glycerol (a "backbone" of carbon, hydrogen, and oxygen). Chemically, this is a triester of glycerol, an ester being the molecule formed from the reaction of the carboxylic acid and an organic alcohol'. Thus, most fats share some chemical features but also are very non-polar and 'fatty'. All organisms contain fatty substances but exactly which one depends on the individual organism. Hence, the different oils discussed previously (olive, sunflower, rape, walnut, grapeseed, safflower, etc.) differ in their spectrum of fatty acids. You can fry your egg in any oil because each mixture shares the same basic physicochemical properties but they have different tastes because your taste receptors use the biomolecular properties of a few individual components to produce flavour information to your brain. In the case of plant pigments, some molecules absorb visible (400–700 nm) and ultraviolet (200–400 nm) light. The wavelength of light absorbed by any one molecular structure is so characteristic that if one can very precisely determine which wavelengths are absorbed by a sample then one can deduce a great deal about the structure of the molecule. Infrared wavelengths are especially well absorbed by a particular type of chemical bonds. Visible parts of the spectrum are absorbed, especially well by certain atoms being grouped in particular ways, for example, multiple conjugated double bonds as found in carotenoids.
9. Lee D. (2007). *Nature's palette—the science of plant colour.* University of Chicago Press, Chicago.
10. Umeno D, Tobias AV, Arnold FH. (2005). Diversifying carotenoid biosynthetic pathways by directed evolution. *Microbiology and Molecular Biology Reviews*, 69, 51–78. See also Schmidt-Dannert C, Cheon Lee P, Mijts BN. (2006). Creating carotenoid diversity in *E. coli* cells using combinatorial and directed evolution strategies. *Phytochemistry Reviews*, 5, 67–74; Misawa N, Satomi Y, Kondo K, Yokoyama A, Kajiwara S, Saito T, Obtani T, Miki W. (1995). Structure and functional analysis of a marine bacterial carotenoid biosynthesis gene cluster and asra xanthin biosynthetic pathway proposed at the gene level. *Journal of Bacteriology*, 177, 6575–84.
11. Canalisation is a term used to describe the progressive limitation of outcomes as a process develops. A simple analogy might be a person drawing. Starting with a single short line on blank paper, the person could start with drawing many objects; he/she could be making a diagram or even producing a letter or number. However, after a second line is drawn, the options for the outcomes decrease. After even a few lines, the person will have been committed to one outcome and there is no going back. In evolution, the sheet of paper representing basic cell biochemistry was already full of detailed lines, dots and shading billions of years ago and the opportunities for further 'improvements' were massively constrained—that was canalisation.
12. Lenser T, Gee DR. (2005). *Modelling the evolution of secondary metabolic pathways.* University of York, MPhil Project Report.

Chapter 10: The Genetic Modification of NP Pathways

1. If one adds a gene coding for Bt toxin (a protein that is very toxic to insects made by some species of bacteria) to a plant, it is predictable that the plant will be more toxic to most insects that eat the plant (although mutants in the insect population will become increasingly resistant to Bt toxin). There is uncertainty as to the consequences of those insects being killed, if the genetically modified crop is grown in the field because ecological interactions are complex—that is why so much effort and energy has been spent arguing as to whether Bt corn or Bt oilseed rape should be grown in Europe. There is little or no uncertainty about the effect of humans or animals eating such crops. First, in the case of grain crops, the gene coding for the Bt toxin could be placed under the control of a promoter which ensures that the Bt gene is only expressed in the leaves or stem and not in the grains that are harvested for use. Furthermore, experiments could be conducted to determine precisely how much Bt toxin needs to be ingested to harm laboratory or farm animals (and by extrapolation humans). If the actual dose that the animals would receive after eating the crop is very much below that level, it could safely be predicted that the crop is 'safe' for farm animal or human consumption. Thus, with enough effort, a 'channel' of predictability can be constructed, linking the insertion of the gene(s) and the consumption of one product by a few populations of organisms. But the sea of uncertainty around that channel should be acknowledged, not ignored. With effort, a logical analysis could reduce the unknowns and confidence about the wider safety issues associated with the growing of the Bt crops could be increased. The past decade has seen the adoption of a number of different Bt crops by an increasing number of farmers in North America, parts of South America and China with no reported problems to consumers of the crops but the long-term effects on the agricultural ecology are still not fully documented. This has given confidence to the advocates of GM crops but the satisfactory experience (to date) of the Bt crops cannot provide a universal lesson. There can be no universal lesson about the wisdom of growing a GM crop because each gene added is unique. In the case of the manipulation of the NP composition, it might be that establishing a 'channel of predictability' will be very difficult due to the inherent unpredictability of NP metabolism.
2. There might be a notorious early example of the unpredictability of manipulating pathways in microbes as evidenced by the presence of toxic contaminants in tryptophan, once sold widely as a 'health supplement'. It has been argued that in the late 1980s, the main supplier of tryptophan, the Japanese company Showa Denko, had adopted a novel, genetically modified strain of bacteria to produce L-tryptophan and their purification failed to remove some new, unexpected minor contaminants which were highly toxic. See Crist WE. *Toxic l-tryptophan: shedding light on a mysterious epidemic*, http://www.seedsofdeception.com/Public/L-tryptophan/1Introduction/index.cfm
3. Lewinsohn E, Gijzen M. (2009). Phytochemical diversity: the sounds of silent metabolism. *Plant Science*, 176, 161–9.
4. Kutchan T. (2005). Predictive metabolic engineering in plants: still full of surprises. *Trends in Biotechnology*, 23, 381–3.
5. Kristensen C, Morant M, Ollsen CE, Ekstrøm CT, Galbraith DW, Møller BL, Bak S. (2005). Metabolic engineering of dhurrin in transgenetic *Arabidopsis* plants with marginal inadvertent effects on metabolome and transcriptome. *Proceedings of the National Academy of Sciences*, 102, 1779–84.
6. Firn RD, Jones CG. (1999). Secondary metabolism and GMOs. *Nature*, 400, 13–14. See also Firn RD. (2006). The genetic manipulation of Natural Product composition—risk assessment when

a system is predictably unpredictable. In K Moch (Ed.), *Epigenetics, transgenic plants and risk assessments* (pp. 20–31). Öko-Institut, Freiburg.

7. Jørgensen K, Rasmussen AV, Morant M, Nielsen AH, Bjarnholt N, Zagrobelny M, Bak S, Møller BL. (2005). Metabolon formation and the metabolic channelling in the biosynthesis of NPs. *Current Opinion in Plant Science*, 8, 280–291.
8. *Equivalence.* Crops grown for human consumption as usually judged on the basis of the equivalence of the processed product to existing products. So, flour made from Bt wheat could be regarded as equivalent to flour made from ordinary wheat. It is possible to aim for equivalence by choosing carefully and targeting where the added genes are expressed. For example, one can control in which tissues a gene is expressed so it would be possible to have the Bt gene silent in the wheat grains themselves so that there would be no Bt protein in the flour (proteins do not usually move between plant organs). One could check out the flour to demonstrate that the Bt toxin level was indeed as near to zero as the analytical method would allow. But what is the equivalence when dealing with NP composition? Are the two very different flavours of apples, Granny Smith and Cox's Orange Pippin, equivalent? Is spearmint equivalent to peppermint? So the concept of equivalence is hard to tie down and is difficult to apply universally.
9. The 'primary metabolites' include such compounds as simple sugars, amino acids, common lipids, the common nitrogen compounds and many phosphorylated compounds.
10. A fine example of this problem is the detection of an NP responsible for one species of insect finding its host plant. An analysis of the volatile chemicals made by the plant showed some chemicals made in large amounts, others in smaller amounts and some in minute amounts. By analysing the nerve output of the insect's olfactory cells to these chemicals, the insect was found to be responding to one of the very minor chemicals.
11. For example, many microbes make rather few NPs when growing rapidly and only start making large amounts of NPs when they enter a much slower growth phase. In plants, insect or fungal attack induces very large changes in NP composition and there is also evidence that plant nutrition, light and temperature can influence NP composition. In studies of the NP composition of a single species, wide variations of NP composition have been found between individuals in the population. In large perennial plants such as trees, the NP composition of leaves, for example, on different branches can reveal significant changes in NP composition.

Index

Note: page numbers in *italics* refer to Figures.